SILVANA DE SOISSONS

NATURAL SKINCARE FOR ALL SEASONS

A MODERN GUIDE TO GROWING
& MAKING PLANT-BASED PRODUCTS

SILVANA DE SOISSONS

NATURAL SKINCARE FOR ALL SEASONS

A MODERN GUIDE TO GROWING & MAKING PLANT-BASED PRODUCTS

PHOTOGRAPHY BY JASON INGRAM

PAVILION

To Mariella, with all my love

First published in the
United Kingdom in 2022 by
Pavilion
43 Great Ormond Street
London
WC1N 3HZ

ISBN 978-1-911663-81-2

A CIP catalogue record for this book is available from the British Library.
10 9 8 7 6 5 4 3 2 1

Reproduction by Rival Colour Ltd., UK
Printed and bound in Italy by L.E.G.O. SpA
www.pavilionbooks.com

Illustrator: Zoe Barker
Photographer: Jason Ingram
Commissioning editor: Sophie Allen
Design manager: Alice Kennedy-Owen

This book is intended as a guide for natural skincare product making at home, for private use and not for the professional skincare product formulator, as there are many laws and tests that need to be complied with if you intend to sell your products to customers.

A note on nut allergies:
The old adage goes that if you can't eat something, don't put it on your skin either. If you suffer from Type I anaphylaxis-type allergies for nuts or seeds such as almond, sunflower, peanut, macadamia or argan, then avoid those oils for skincare products. Coconut, safflower and sesame oil-based products might be more suitable for your skin.

A note on the use of essential oils during pregnancy:
In the formulations in this book we often use essential oils to derive the active benefits of the plant's properties. When used topically, a few drops of essential oils are safely diluted in carrier plant oils before being applied to the skin. They are never taken internally or applied undiluted. During pregnancy, however, you may wish to check with your doctor before using essential oils.

A note on production lead times:
Certain botanical skincare products take a few weeks in the making – for example, plant material may need to be steeped for a number of weeks in the making of oil macerations or tinctures. Always make sure that you read the formulations to the end, so that you know if an ingredient takes a long time to make. In my workshop I have a number of tinctures, botanical oils and hydrosols pre-prepared, to enable me to accomplish a wide range of products.

CONTENTS

AUTHOR'S NOTE

NO MAN FEARS WHAT HE HAS SEEN GROW
Old African proverb

This book is a distillation of all that I have learnt from a life of growing plants and from their applications: first from my time in retail and then from my own soap-making business in Dorset – Farm Soap Co. – my commercial garden, herbal foraging and skincare workshop. I wanted to bring to one cohesive whole a simple, modern guide for creating your own seasonal skincare products.

Grown or gathered and made by you for yourself and your family, you will know exactly what ingredients are used in these products and can benefit from their healing and nurturing properties in the knowledge that they are good for the planet as well as for people.

This book puts YOU, the reader, grower and maker, in control.

Home-grown and foraged plants are full of vitality, colour and goodness, with nutrients that can help support your skin's natural microbiome. Making small-batch, fresh products ensures you avoid chemical preservatives, additives, colourants and fragrances and can embrace the seasonality and richness of nature's offerings – and often for a fraction of the price of commercial ranges.

Healing, calming and entirely natural, every formulation in this book has been carefully researched to help you detoxify your bathroom cabinet and skincare routine, and replenish it only with a few fresh, handmade products that are powered by plants, not manufactured chemicals.

Wherever you can grow plants – be it in your own garden or allotment, on a balcony, patio or a few pots on a window sill – I know that they can bring physical, emotional and mental well-being, while keeping you fit and fascinated by the natural world. My hope is that you will have as many years of enjoyment growing, making and using these natural skincare recipes and other plant-based products for your home as I have.

Silvana

1

AN INTRODUCTION TO THE SKIN

The skin is our body's largest organ and one of its most self-sufficient. It's a highly effective barrier, protecting us from pathogens, pollutants and dehydration, expelling toxins while simultaneously absorbing nutrients – including the manufacture of vitamin D from sunlight – and helping to regulate our temperature. It's truly a multi-tasking titan.

Skin acts as a sensory organ – it allows us to feel touch and pressure, temperature and pain. These perceptions begin as signals generated by touch receptors and travel along sensory nerves that connect to neurons in the spinal cord. Signals then move to the thalamus, which relays information to the rest of the brain; we can feel warmth or cold, pleasure and arousal or hurt and discomfort, all through the skin.

Remarkably, skin is able to self-heal, self-regulate, self-cleanse and self-moisturize. It even has its own microbiome, a community of beneficial microorganisms – yeasts, bacteria and fungi – that acts as a biological barrier to protect us efficiently from harmful pathogens and feeds on our sweat and dead skin cells. This microbiome is vital because it forms part of our immune system, connected to the gut and the brain, and helps to reinforce the skin's role as a barrier against infection.

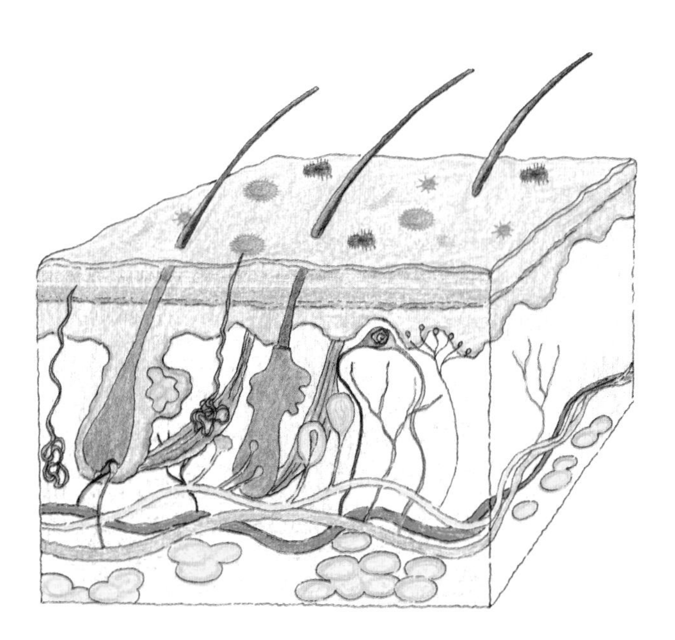

THE LAYERS OF THE SKIN

Our skin is composed of three layers. The outermost, the epidermis, is virtually waterproof thanks to the stratum corneum, the visible surface that also gives us our skin tone. Melanocytes, the cells that protect us from UV radiation, are present in the stratum corneum and they create the pigment melanin as a natural form of self-protection, which is why our skin darkens when exposed to the sun.

The acid mantle is a thin, invisible layer on the surface of the stratum corneum which is made from a natural oil known as sebum, secreted by the sebaceous glands. It protects the skin from damage, keeps out dirt and impurities, and is an important barrier in preventing dehydration, keeping our skin resilient and glowing naturally. We need to support the acid mantle by avoiding harsh cleansing products or exfoliators, over-exposure to the sun, and irritants such as pollution and other contaminants. The sweat secreted from our skin is another form of defence in the mantle, as well as a temperature-regulating mechanism. The pH of the skin is normally quite acidic, around 5.5, whereas the pH of the blood is more alkaline, around 7.5. This difference in pH helps to prevent bacteria reaching internal tissues.

Below the stratum corneum the epidermis itself is largely made up of protein cells called keratinocytes. These continually move to the skin's surface to replace the older, visible ones that are sloughed off in a two-week cycle – a natural form of cleansing, exfoliation and tissue regeneration. The skin's natural fats or lipids are located in the epidermis. They maintain the strength of the acid mantle. They also aid the skin's natural repair process.

Located in the dermis, the middle layer of the skin, are hair follicles and sweat glands, both of which are self-regulating temperature controllers, and sebaceous glands that produce sebum to help us self-moisturize.

Finally, the hypodermis or subcutaneous tissue is the innermost layer of our skin, which is made up of fat and connective tissue.

SAVE OUR SKIN

IN THE FACTORY WE MAKE COSMETICS. IN THE STORE WE SELL HOPE.
Charles Revlon (1906–75)

The ability of our skin to regenerate and recover is unlike that of any other organ in the human body. Yet, global brands seek to convince us that our skin is constantly under threat, from dirt, pollution, the weather and ageing, and they offer a myriad of essential, ever-changing and evolving beauty products for every part of our skin – literally from scalp to toe.

Modern-day beauty standards apply unrealistic and unsustainable pressure, on women in particular, to keep consuming, meaning large pharmaceutical and cosmetic companies can keep profiteering.

The global skincare industry is worth hundreds of billions of dollars. Even the self-styled natural and organic skincare sector is valued in billions, although it is debatable what is considered 'natural' as a great deal of 'greenwashing' is applied to the industry. This enables a positive spin and public-relations strategies that lead consumers to believe that a product, process or brand is far more environmentally friendly than it actually is.

The use of many of the words appearing in the sales and marketing literature of toiletries and cosmetics is largely unregulated and far from transparent – the terms 'plant-based', 'natural', 'botanical' and 'gentle', for example, can be used freely to convince us that what we are buying is good for us. Packaging can feature illustrations of bees or flowers and persuasive names that evoke organic gardens, artisan processes or a tropical paradise, even if the product itself contains harmful derivatives from the petrochemical industry, among them parabens, synthetic fragrances, phthalates, triclosan, sodium lauryl sulphate, mineral oil, silicone, propylene glycol, and ethanolamine, that build up in our bodies. The word 'fragrance' on a product label can refer to a combination of up to 200 individual synthetic fragrances, and there is no requirement to list individual fragrances.

BEAUTY AT ANY PRICE?

One estimate suggests that each year the average woman in Britain is absorbing up to 2.2 kilograms (almost 5 pounds) of toxins from the 'cocktail' of chemicals applied to the skin through body wash, sun screen, shampoo, conditioner, masks, deodorant, creams, make up, perfume and many other toiletries and cosmetics.

It's not simply that the use of toxic chemicals in personal care products affects us as individual consumers, they are entering our food chains and polluting the environment. One class of toxic fluorinated substances, nicknamed the 'forever' chemicals – the microbeads and microplastics used as exfoliants in place of natural ingredients in some cosmetics and toiletries including toothpastes – enter the water systems, contaminating soils and seas and ultimately threatening vulnerable marine flora and fauna, including coral reefs. Legislation banning these chemicals has now been passed in many countries because of their lasting impact.

Other cosmetics and perfumes directly impact on species because they contain animal-derived ingredients – these include powder from silkworms, ambergris from whales and squalene obtained from shark liver oil – which are then tested on laboratory animals before being passed as safe for humans. We need to stop supporting the companies making their fortunes with such disregard for the world we all share.

Women still tend to be judged by their looks and feel pressured to meet expectations, which is fuelled by skincare and cosmetic conglomerates keen to persuade them to buy multiple products for different parts of the body according to skin type and age. The vast PR and marketing budgets of these conglomerates finance free products, fees, sponsorships and collaborations for and with social-media influencers, beauty editors, journalists and, of course, celebrities in exchange for positive endorsements and reviews. Their viral power as tastemakers in a crowded and competitive market is compelling. Instagrammers can monetize their large followings by the use of photo feeds, reels and stories, showcasing their daily skincare rituals and make-up applications. It's time for women to close their eyes to the marketing hype and instead focus on caring naturally for their skin.

THE REALITIES OF SKIN AGEING

As we age, collagen and elastin – the proteins that provide skin structure and elasticity – are depleted and skin starts to line, wrinkle and sag, becoming more fragile, translucent and thinner. Time, stress, our lifestyle patterns and even gravity take their unforgiving toll – inevitably leading many consumers into the hugely profitable arena of anti-ageing skincare. It's a low-hanging fruit – we all want to age well and are easily led down the rabbit hole in pursuit of tight and plump skin.

Despite the pseudo-science and empty promises of eternal youth, there are no skin creams or lotions that can turn back the clock; instead medical research points repeatedly to the basics of good skincare and lifestyle practices – frugal, simple and within our reach – practices that complement and support the self-sufficiency of our skin's microbiome, not replace it.

In all the natural skincare research I have undertaken, the golden rules for looking after your skin from the teenage years to older age focus on good habits and natural care. Just like the food we eat, once you know the ingredients and the source, you are better informed to make wise choices, and common sense is hugely important.

BATHROOM KIT FOR EVERYDAY USE

Linen cloths – These are the most important tools in your skincare armoury, used as face flannels for cleaning skin and exfoliating without the need for pastes or scrubs that contain microbeads, and for applying and removing product, thereby saving on cotton wool which is not environmentally friendly. I wash my used linen cloths daily or every other day in my normal wash cycle at 30°C (86°F) with eco-friendly washing powder.

Body brushes, loofahs and bath mitts – These brushes are really invigorating! Try to buy a natural bristle brush and use it to massage your skin, working upwards, towards the heart, helping to bring oxygenated blood to the skin every morning or evening, before a bath or shower. Loofahs and bath mitts are also extremely useful for dry-brushing the skin before a shower – I use mine nearly every day, from head to toe. I wash them regularly in hot soapy water to prevent the accumulation of dead skin cells or bacteria.

Wool or cotton face pads – As a sustainable option, you can make reusable wool or cotton crochet or knitted squares or circles instead of using single-use cotton wool pads, which can also be used around the sensitive eye area. (Even the production of organic cotton wool is not environmentally friendly.) You can also cut up old towels to use in this way – a drinking glass or cookie cutter makes a good template, then sew round the edges to prevent any fraying.

Pumice stone – Pumice is formed when hot volcanic lava meets cool water. It is a light but mildly abrasive stone – perfect for the removal of dead skin on cracked heels, elbows or knees.

TEN PILLARS OF NATURAL SKINCARE

Over the years I have whittled down the essentials of caring for skin without recourse to chemicals, unethical products or expensive treatments. The following sections provide the foundation of my philosophy for a healthy life and beautiful skin.

PLANT-BASED PRODUCTS

Using gentle plant oils and extracts rather than synthetic petrochemical derivatives assures us better-quality, nutrient-dense cleansing and moisturizing. Our skin cells do not recognize the chemicals used in many industrially manufactured skincare products and cosmetics, and the immune response is to trigger T cells, a type of white blood cell, to ward off these foreign substances in the same way they fight diseases and pathogens. Chemicals displace the natural lipids in the epidermis, clogging the pores and irritating the skin, which is how dermatitis and allergic rashes occur. In contrast, natural oils and other plant-based ingredients are easily absorbed and help to nourish the skin and support the microbiome without stripping or harming the acid mantle.

SKIN MASSAGE AND BRUSHING

All-over body massage and dry brushing are known to promote healthy skin. Massaging helps to improve circulation, enabling the lymphatic and blood systems to bring infection-fighting antibodies and nutrients to the skin, making you feel pampered and refreshed.

These simple friction techniques slough off dead skin cells and help remove impurities through the pores. Acne occurs when the pores of the skin become blocked with oil, dead skin or bacteria, and can take hold at any age, which is why it's important to make these techniques part of your skincare routine.

1. Dry skin brushing is done before a shower or bath using a gentle, natural bristle brush. Stand naked on a clean towel in a warm room.
2. Starting from the tips of your toes, gently brush your skin all over with circular, upward motions in the direction of the lymph flow, towards your heart. It should feel pleasant, not painful so keep the pressure light.
3. Pay particular attention to areas of skin that need extra circulation flow, like your heels, feet, knees and elbows.
4. Gently brush each part of your body several times, until you feel a tingly, warm sensation. Don't spend more than say 5 minutes in total.
5. Enjoy a warm shower or bath, dry your skin then rub a botanical oil or balm all over.

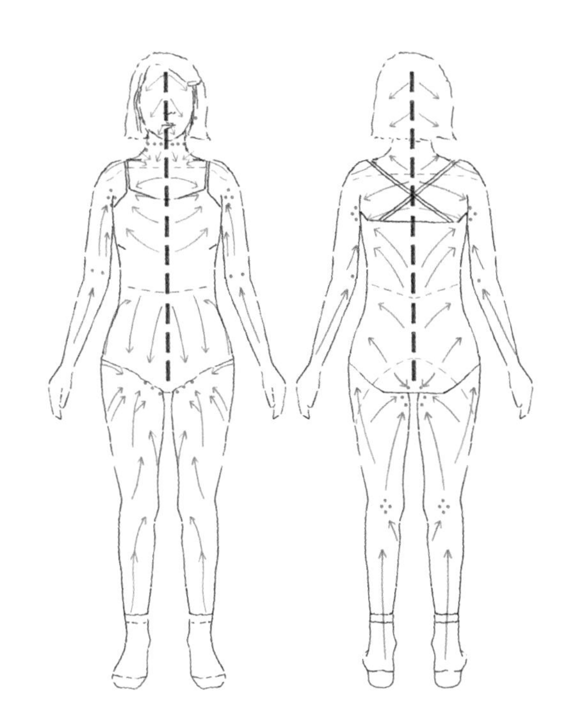

A facial massage before bedtime, using a natural skincare oil, also improves the tone and health of our skin and is the ideal way to moisturize overnight. Any excess oil can be removed by applying a dry flannel or linen cloth to the face and pressing lightly.

1. Cleanse your face and hands.
2. Warm some botanical oil between your palms (like almond oil, olive oil, rapeseed oil, coconut oil, rosehip oil or borage seed oil).
3. Apply the oil to your face, avoiding your eyes, using a circular action and gentle pressure.
4. Massage the lymph areas below your ears, working outwards and upwards along the sides of your neck, both front and back.
5. Gently pat your face with your fingertips, smoothing lines and encouraging blood circulation.
6. Very gently press under your eyes with your fingertips then slide them out towards your temples.
7. Rub any excess oil into the backs of your hands.

SLEEP

An undisturbed 7–8 hours of restful sleep every night is vital for good skin. Sleep 'hygiene' is therefore very important to ensure you wind down. An hour or so before bedtime, try to relax by drinking herbal tea, taking a warm bath, dimming the lights or reading a good book. Rather than stimulate your brain by looking at a TV screen or digital devices – instead you could try meditation or prayer to help prepare you for rest. During sleep, the skin's blood flow increases, and the organ rebuilds its collagen, repairing damage from UV exposure, reducing wrinkles and age spots. Sleep does far more than restore our mood or energy levels; it boosts our immune system and heart health, prevents inflammation and enhances our productivity.

CALM

So-called oxidative stress is a major cause of skin problems and ageing. We of course rely on oxygen, but normal metabolic processes sometimes cause paired oxygen molecules to split into two single atoms with unpaired electrons known as free radicals. Long term, unless the free radicals are balanced by antioxidants in the body, oxidative stress can damage cell membranes and proteins and DNA.

Among the cells that are damaged are those responsible for the production of collagen, elastin and hyaluronic acid that give youthful skin its firm structure. Production of these substances naturally declines as we get older and oxidative stress accelerates this process.

Free-radical production is both triggered and increased by mental stress, pollution, UV exposure, poor diet, smoking and alcohol consumption, so pursuing a more mindful, calmer life, eating foods that are rich in antioxidants and making better lifestyle choices can in turn improve our skin.

WATER

From birth, more than 50 per cent of human body weight is water, a percentage that is maintained into adulthood but which declines as we age, more so for women. Drinking water daily is vital to maintain body temperature, flush out toxins and help our brains to function. It keeps our skin hydrated and prevents skin disorders and premature wrinkles. Most experts believe the optimum amount of water to drink every day is eight large glasses, or roughly 2 litres (3½ pints).

If that sounds a lot, remember your intake can be interspersed throughout the day and made to taste more appetizing as herbal teas (see page 108) or by adding some zing with slices of fresh lemon, lime, cucumber, apple or herbs such as mint, rosemary, fennel or lavender. In winter months, hot water flavoured with lemon, grated ginger and honey left to steep for 2–3 minutes before being strained into a cup makes a warming and hydrating drink.

GREENS

Eating a varied, balanced diet that includes a wide range of multi-coloured fruits and vegetables as well as nuts, seeds, grains and pulses benefits the whole body and consequently, the skin as well. Proteins – in particular the amino acids found in eggs, fish, dairy and chicken – are essential tissue and collagen building blocks while green leafy vegetables are especially beneficial to our skin because they contain the green pigment chlorophyll, a powerful anti-inflammatory that helps with conditions such as acne, eczema and rosacea. Kale, lettuces, cavolo nero, microgreens, spinach and leafy herbs are high in antioxidants and phytonutrients: vitamin A, for helping your body's immune system; vitamin C for healthy skin (and improved collagen production) and bones; and vitamin K, which is anti-inflammatory, antioxidant and collagen promoting, helping to heal summer skin from over-exposure to sunlight.

Green vegetables are also a rich source of minerals and trace elements: potassium, which is a water regulator for the skin; magnesium, which helps with its elasticity; and iron, which aids skin healing. The high content of balanced Omega-6 and Omega-3 alpha linolenic fatty acids found in greens are important components of skin cells, without which the skin would have a dry and aged appearance.

The folic acid and beta-carotene found in whole grains, fruit and vegetables also help to prevent sun damage. Folate, also known as vitamin B9, is vital for

the synthesis of amino acids, the protein building blocks we need to keep skin plump, especially in later years, when collagen production stops.

Finally, green vegetables have a high water content and hydration is key to healthy skin.

SHADE

Exposure to UV light accelerates the changes to our skin as we age and can cause skin cancers. UVB and particularly UVA rays damage collagen fibres, resulting in dry, leathery skin. Age spots (hyperpigmention), often seen on the back of hands which are exposed to sunlight during outdoor activity or driving, are patches of brown skin which is the pigment melanin's way of protecting the skin by darkening it. One of the best ways to protect your skin in the summer is not by slathering chemical sunscreen filters or after-sun products, but instead by wearing clothes and a wide-brimmed hat to cover it, and preferably by staying in the shade. If you have pale skin, play safe: 10 minutes a day in the sun is enough to create vitamin D, an important vitamin that the cholesterol in our skin can synthesize when it is exposed to sunlight; only if you have naturally darker skin should you contemplate spending longer in the full heat of the summer sun.

FRESH AIR AND GENTLE EXERCISE

Walking, swimming, yoga, dancing and other forms of gentle exercise benefit the skin because by raising the heart rate and increasing the blood flow round the body, more oxygen and nutrients are carried to working cells all over the skin. The increased levels of sweat flush out more toxins through the skin pores, alleviating pressure on the kidneys and liver. Fresh air and exercise during the day are known to support the skin's microbiome and aid sleep at night, both of which are also crucial.

CLEAN BEDDING

On average, we spend about a third of our lifetime in bed, which is why good bedding hygiene is essential to ensure that our skin remains clean during its most regenerative part of the day – the time we spend asleep.

Even if you bathe before bed, as you sleep your bedding and night clothes will accumulate traces of sebum, dust, pollution, bacteria, bodily fluids, make-up, hair products, perfume – even faeces from dust mites that reside in the mattress, duvet, blankets or pillows themselves. Washing and airing bedlinen, turning and vacuuming mattresses and changing pillows frequently will all benefit your skin, as well as increasing your chances of a good night's sleep. Your face rests on pillowcases so they in particular need to be washed regularly, preferably twice a week.

CLEAN COMBS, SPONGES AND BRUSHES

Make sure that anything that touches your skin or scalp is clean and fresh. A build-up of dead skin cells, bacteria, old make up and sweat can accumulate on all make-up and hair brushes, combs and sponges – so once a week wash them with warm water, ecological liquid soap and vinegar. Dry them off with a clean towel and leave to dry fully in warm air on a sunny window sill or outside.

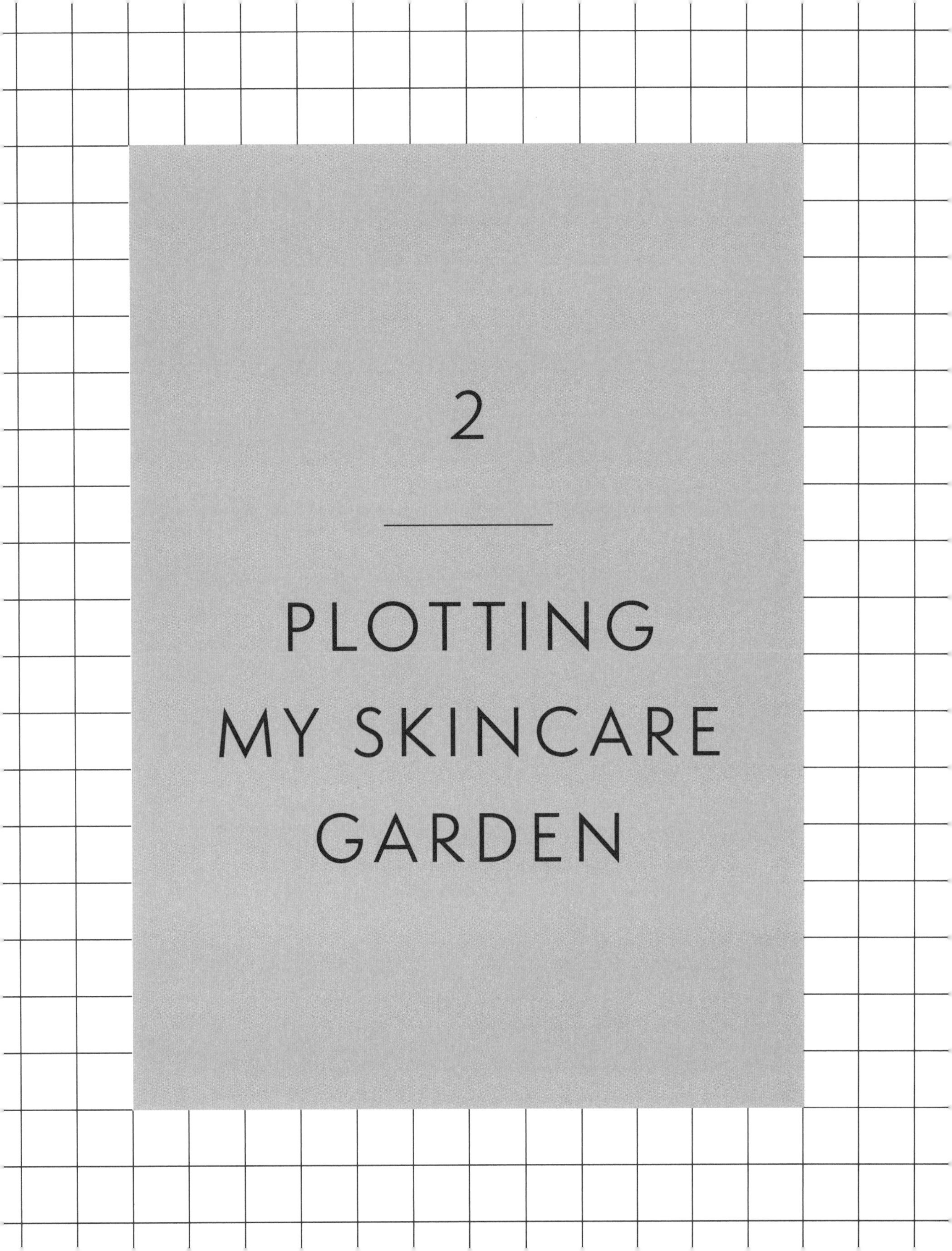

2

PLOTTING MY SKINCARE GARDEN

IN THE BEGINNING

The courtyard garden and entrance next to my workshop in Dorset is smaller than four parking spaces, while the walled allotment garden is about three-quarters the size of a tennis court, but both are densely planted and every possible surface, corner and vessel has been utilized, from beds to baskets to boxes.

The courtyard faces south and has useful spaces along its walls, where I've positioned hanging baskets, trellis and wall containers. None of these need to be expensive and they are useful for growing plants that thrive with wall protection, like Mediterranean herbs, or vigorous ones, like mint, that tend to take over beds if left to their own devices. Many of the plants grown for skincare are especially tactile and fragrant, so whenever I make an initial plan, I think about how the space will be used and position tubs of rosemary and thyme next to a door, hanging baskets with mint on a garden fence or wall and I plant rows of lavender down the path so that I can rub the leaves and catch the scent as I walk around.

I am able to grow a very wide range of skincare plants in a small space, for retail, wholesale and private label (branded goods for companies) and, for this book, I feel it's interesting to show how, even in a small space, you too can grow an abundance of plants to use for making products. Space should not limit your ambitions, because herbs and flowers growing in a compact space compete for light, air and nutrients and often yield far better fragrance and essential oils, rather than voluminous, lush foliage.

When I took on the lease of the workshop and the allotment, the plots had been neglected for a few years, so over that first winter, weeding and cutting back the overgrowth was my initial task. I cut all the plant material into small pieces to turn into compost, ensuring that weed seeds were removed and binned.

I then gave the existing soil a good raking over to remove large stones and existing weed roots and took out plants I didn't want – but I didn't dig the soil. I left smaller pebbles as they help drainage and, in early spring, they also keep the soil warmer on sunny days, aiding germination rates. Many herbs and flowers used for skincare are plants originating from regions with warmer climates and poorer soils than my corner of southwest England, such as the Mediterranean and North Africa, so I needed to ensure good drainage and to avoid overfeeding the soil. Above all, it's important to try and avoid disrupting the existing structure and microbiome of the soil.

That first winter, I covered the bare earth of my beds with a thick layer of cardboard – old removal boxes or packaging boxes are ideal – to stop light from reaching the entire surface thus suppressing any remaining weeds underneath. The cardboard biodegrades easily and then becomes part of the soil structure. For the bigger beds on the allotment, I used a roll of black biodegradable film (it's made from cornstarch and can be bought online); I then covered this with cardboard. It's not pretty to look at initially, but all you need is patience. Read on!

I spread a 10cm (4-inch) layer of compost over the cardboard, rimmed with bed edging or large stones so that the compost didn't fall into the

9A

path. Using the back of my rake, I firmed down the compost layer so that it settled and covered the entire surface generously.

At this stage, the bare beds didn't look attractive but the cardboard rots down fast and I knew that, come late spring and summer, when the plants would be growing rapidly, I wouldn't even see the soil. By planting intensively, leaving as little room as possible for weeds to grow, this no-dig method ensures more abundance in the present and less work in the future.

THE PLANTING PLAN

As you can see from my original planting plan, several practical considerations need to be taken into account.

- Proximity to water – including one or two inexpensive water butts to harvest rainwater from the house or a shed roof is a worthwhile idea, for ease of watering in a hot summer and to avoid using tap water, especially if it is chlorinated.
- Space for a compost heap container – this can be made from inexpensive pallets, or, if you have room, you can just make a compost 'cake'. (See chapter 4, page 74, for what it takes to make compost.)
- A bin for waste – weed seeds need to be removed from the garden and be separated from compost material or they will take over, so it's important to get a garden waste bin, or, if you can, create a small, safe bonfire area at the bottom of the garden.
- Space for small and larger tools to be stored safely and protected from the elements.
- A table and chair so that you can sit comfortably with a cup of tea and enjoy the fruits of your labours.

The list of plants that can be grown for skincare products is potentially very long. In the next chapter, I have selected 17 that have effective impact in terms of their antioxidant, anti-inflammatory, antimicrobial and moisturizing properties. You may not be able to grow every one if space is at a premium – just select a few to fill the size of your plot, containers, balcony or window sill, also bearing in mind how much time you have available to look after the garden. In addition, there are plants that can be foraged, and so are available to everyone.

NO DIG GARDENING

The No Dig Gardening movement is growing rapidly across the world, with many acolytes who can swear by the success and simplicity of the method. One of its most prominent ambassadors, Charles Dowding, runs very successful garden courses from his commercial plot in Somerset and his videos are widely followed on YouTube and social media.

By not digging and disturbing the soil structure, the complex and interdependent communities of millions of bacteria, fungi, nematodes and microbes that enrich and aerate the ground are left to create nutrient-filled and well-drained soil that is the perfect medium in which plant roots can develop and grow, uncompacted, without the need to add artificial fertilizers. As a bonus, annual weed seeds often germinate when the soil is disturbed through digging – by leaving the soil structure undisturbed you are saving yourself a great deal of hoeing later on in the year.

The key to No Dig Gardening is to add a generous layer of 5–8cm (2–3¼) microbe-rich compost to the surface every year, which helps to suppress any weeds, keeps in moisture and adds nutrients to the soil.

You can switch to a no-dig system at any time and gradually the soil microbiome will develop as composts and mulches are added.

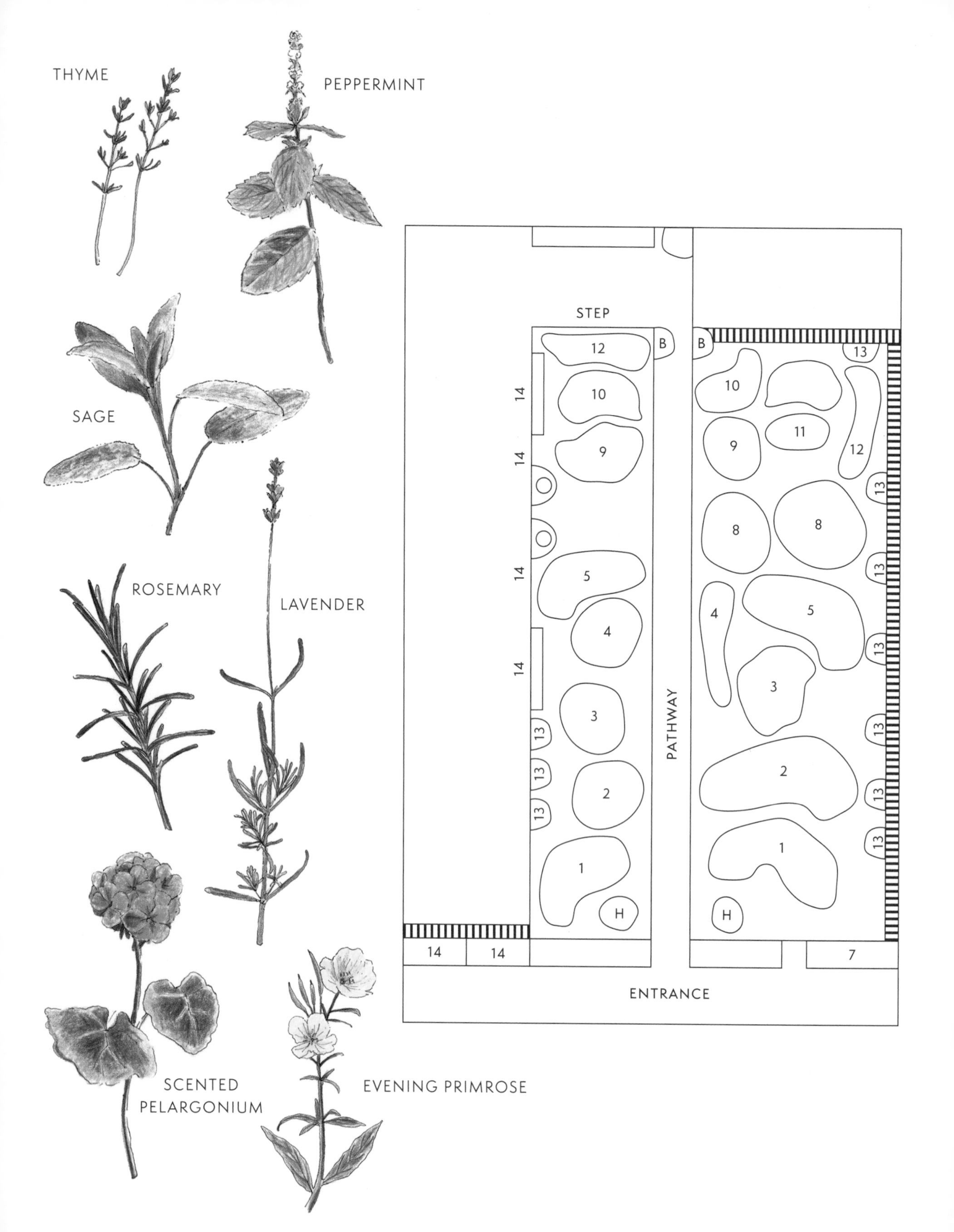

THYME
PEPPERMINT
SAGE
ROSEMARY
LAVENDER
SCENTED PELARGONIUM
EVENING PRIMROSE
STEP
PATHWAY
ENTRANCE
B
B
H
H
1
2
3
4
5
8
8
9
9
10
10
11
12
12
13
14
7

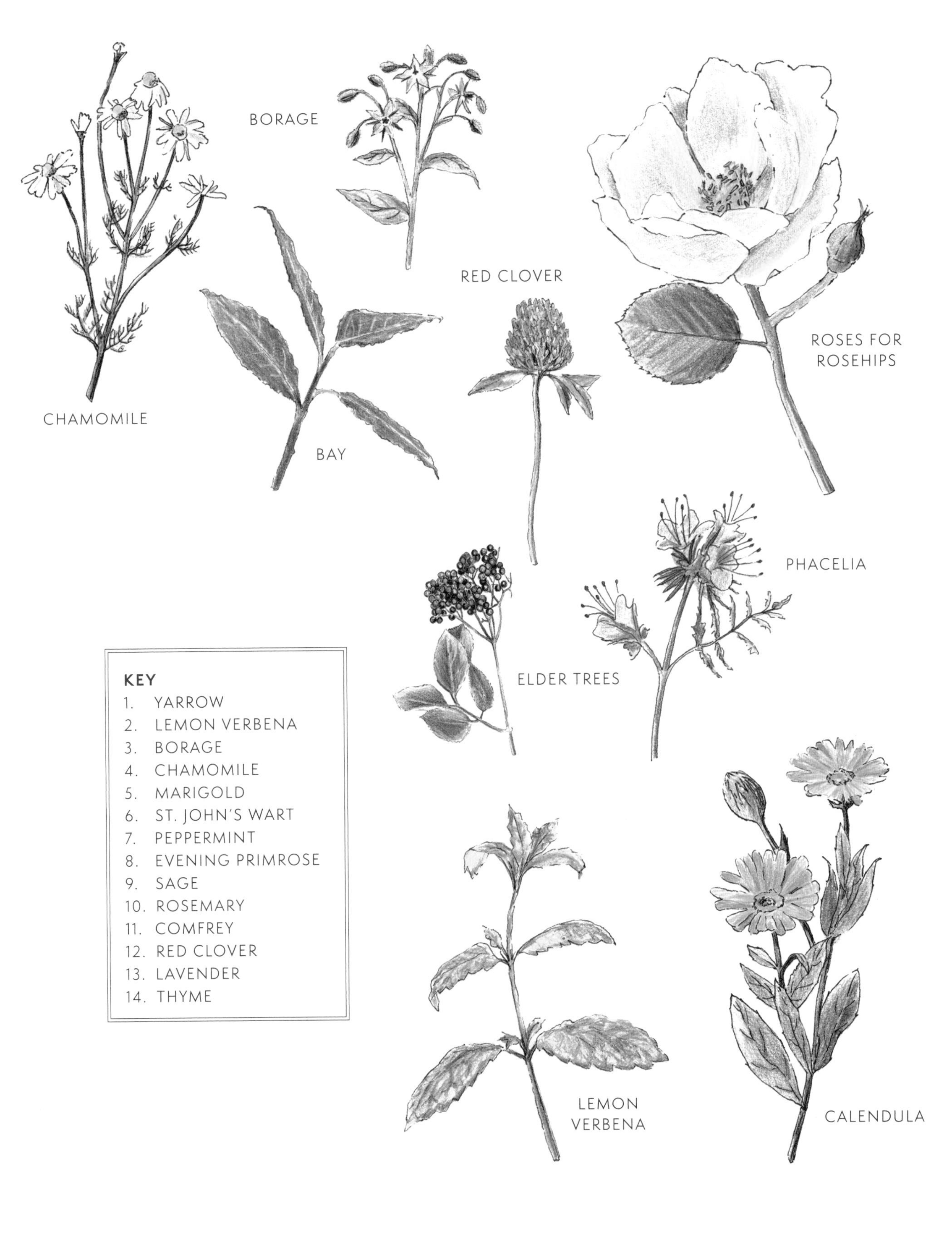
BORAGE
CHAMOMILE
RED CLOVER
ROSES FOR ROSEHIPS
BAY
PHACELIA
ELDER TREES
KEY
1. YARROW
2. LEMON VERBENA
3. BORAGE
4. CHAMOMILE
5. MARIGOLD
6. ST. JOHN'S WART
7. PEPPERMINT
8. EVENING PRIMROSE
9. SAGE
10. ROSEMARY
11. COMFREY
12. RED CLOVER
13. LAVENDER
14. THYME
LEMON VERBENA
CALENDULA

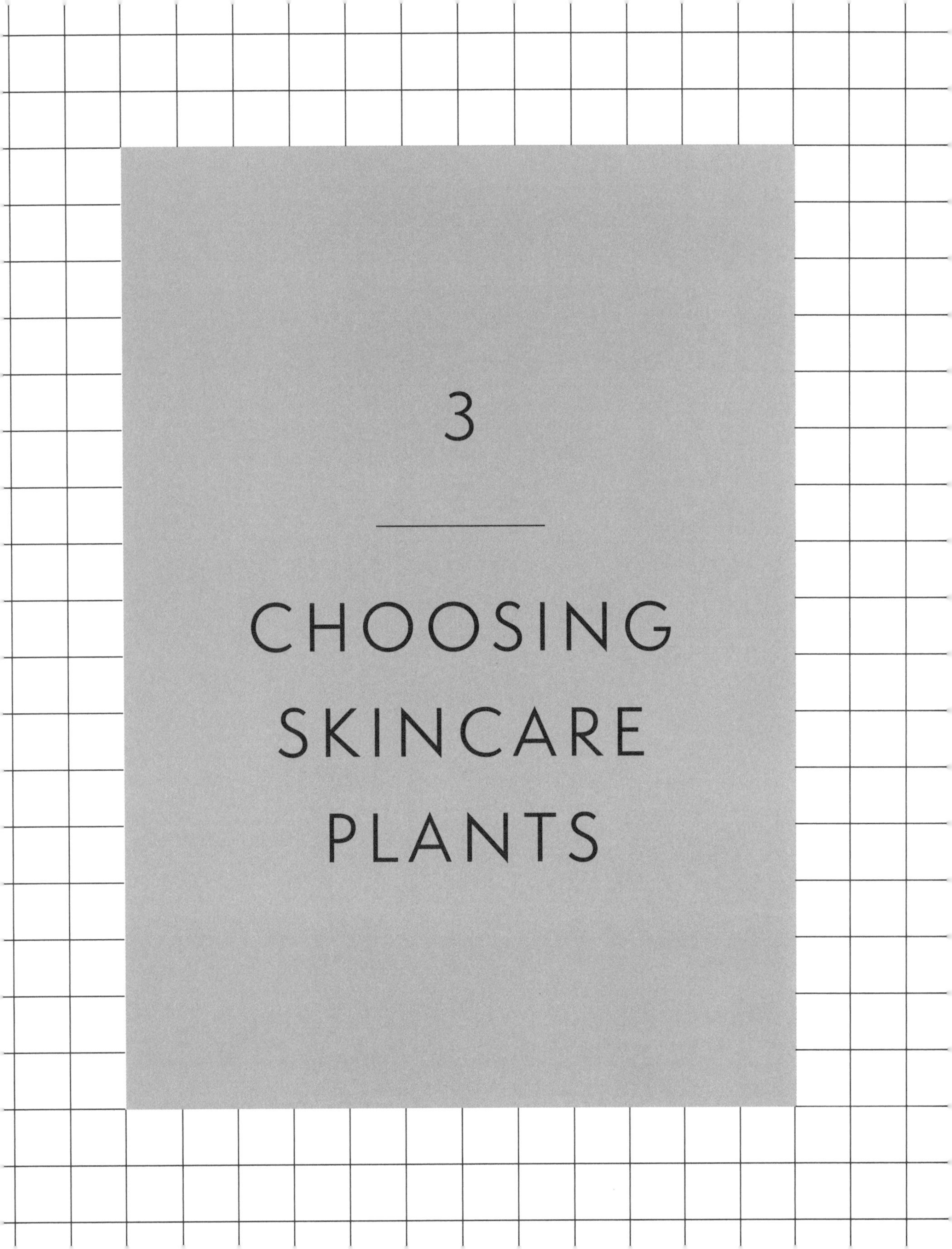

3

CHOOSING SKINCARE PLANTS

When deciding which skincare plants to cultivate, the main considerations are: the size of your space – balcony, courtyard, allotment or garden, the aspect – the cultivated skincare plants in this book are mainly Mediterranean plants so they need sun and good drainage, how much time you have to dedicate to it and which plants are high in antioxidant, anti-inflammatory, antimicrobial and antiseptic properties for cleansing, healing and moisturizing the skin.

The plants I profile in this chapter have been used for making natural skincare products for millennia, by ancient and indigenous civilizations from the Romans to the Greeks, Egyptians, native North and South American populations, as well as by Chinese and Ayurveda traditional medicine practitioners, medieval apothecarists and Victorian herbalists.

Long before the advent of the modern pharmaceutical industry, people had learnt which plants could make poultices, compresses, lotions, infusions and oils to treat their skin for reasons of both health and beauty. In ancient Greece, the physician Hippocrates, who became known as the father of modern medicine, treated skin ailments with herbs and resins.

Botanically, a herb is a plant that has fleshy rather than woody stems – hence the term herbaceous – and many of the herbs historically used for healing can be identified today from their Latin species name, *'officinalis'*, which indicates plants used for medicine, herbalism or cooking. These herbs have active medicinal principles, either toxic or nontoxic. The latter include bitterness, essential oils, mineral salts and tannins, each with different properties:

- **Bitter herbs** – tonics that stimulate appetite.
- **Essential oils** – extracts that can be germicidal, antioxidant, antibacterial or anti-inflammatory.
- **Mineral salts** – micronutrients that aid resistance to disease.
- **Tannins** – substances in plant tissues that are astringent and aid healing.

The following 17 plants would have been commonly found in a medieval apothecary garden and provide a useful base to start a skincare garden. In fact, many of these plants were introduced to Britain by the Romans, meaning they may have been cultivated here as food or medicines since AD 43. I have included a further five plants that grow wild and can be responsibly foraged in open countryside. Across the globe, though, literally hundreds of different plants are used for natural skincare. Certain plants come in many varieties – roses, for example, that yield rose-hips for face oils – so you may find yourself planting different varieties, year on year.

RIGHT PLANT – RIGHT PLACE

Always remember that for a plant to thrive it must be in the right place, meaning the right conditions for its particular needs. By considering where plants originate from, you can provide the right habitat for them – for example, good drainage for all the Mediterranean herbs, heaps of mulch and manure for roses, plenty of sunshine for pot marigolds and a little shade for mint. Above all, don't kill with kindness – many of these plants are very low-maintenance.

ACHILLEA MILLEFOLIUM – YARROW

Once used to treat battle wounds, the white musk-scented flowers and feathery leaves of yarrow make an astringent and antibacterial tonic that helps to cleanse and tone the skin. It can also be prepared as a herbal tea.

Yarrow grows wild across the northern hemisphere and can be found in hedgerows and at the margins of fields. The plant grows easily from seeds sown in prepared ground in early spring. It prefers a well-drained position in full sun where it will reach 1 metre (more than 3 feet) in height and 20cm (8in) in width and it is quite invasive.

BORAGO OFFICINALIS – BORAGE

Originally a native of the Mediterranean, the bright blue flowers and hairy edible leaves of the plant, which are rich in vitamins and minerals, have been used for millennia to help combat stress and skin irritations. High in gamma-linolenic acids, Omega-6 fatty acids, borage seed oil is both highly moisturizing and anti-inflammatory, and is widely used in the skincare industry.

Borage grows up to 1 metre (more than 3 feet) from seed sown in well-drained soil in spring and self-seeds enthusiastically in late summer, so space out the seeds when sowing. Its star-shaped flowers attract bees for their nectar.

CALENDULA OFFICINALIS – POT MARIGOLD

One of the most popular Mediterranean medicinal plants with a long history of cultivation, calendula helps to stimulate collagen production. It is widely used in ointment creams for minor cuts or grazes, inflamed skin and acne because it has antioxidant, antiseptic, antibacterial and anti-inflammatory properties. It also helps to regulate sebaceous oil production. Oil macerated with the orange flower heads is highly prized for making cold-process soaps (see page 130 for the calendula, peppermint and poppy seed soap), body and bath oil and creams.

Calendula grows happily in containers or in well-drained, south-facing beds. It's a prolific self-seeder, but do collect as many seeds as possible and dry them to ensure next year's crop, because self-sown seedlings that germinate in late summer or autumn will not survive winter temperatures. As with many cut-and-come-again annuals, the more flowers that are harvested, the more the plant keeps producing.

CHAMAEMELUM NOBILE – ROMAN CHAMOMILE

As its common name suggests, chamomile has been cultivated since Roman times. Chamomile tea was used by the Egyptians, Greeks and Romans as an ointment to treat wounds and promote healing – and its soporific properties mean it is suitable as a herbal tea to help relax us into sleep. With anti-inflammatory and antioxidant properties, the flowers help to soothe skin irritations like eczema, psoriasis and rosacea, reducing redness and blemishes.

Roman chamomile is a small perennial plant that spreads to about 30cm (12in) wide and grows up to 50cm (20in) tall. It has a distinctive earthy, grassy fragrance, which helps to differentiate it from other members of the daisy family.

HYPERICUM PERFORATUM – ST JOHN'S WORT

Another common perennial, St John's Wort has been used to treat wounds and as a tonic for the nervous system since at least the times of the Crusades. Astringent and anti-inflammatory, the herb makes an effective skin-repairing tonic. The oil from its distinctive small, perforated leaves can be extracted and used as a topical treatment for burns. When the star-shaped yellow flowers are fully opened in summer, they can be harvested and macerated in oil to make an anti-inflammatory balm.

The seeds can be sown in autumn in seed trays and planted out in rows in spring. St John's Wort grows best in a sunny, well-drained position and can reach 1 metre (more than 3 feet) in height and width. It is often used as a low hedge.

NOTE OF CAUTION

*Avoid St John's Wort during pregnancy because more research is needed to confirm the safety of the herb to the unborn foetus.

LAVANDULA ANGUSTIFOLIA – LAVENDER

The name lavender derives from *lavare*, the Latin verb to wash, and the herb has been used in perfumery and herbal medicine for centuries. Soothing and sedative, the essential oil is used for respiratory problems, muscle pain and tension, as well as for its antibacterial properties. The flower heads can be infused to make a calming tea, to promote good sleep.

Originally from the Mediterranean, above all lavender needs a sheltered position, well-drained soil and full sun. It dislikes being waterlogged, so it's important to incorporate horticultural grit in the compost mix. The plant is best grown from cuttings or simply purchased in pots. It is evergreen and can reach 90cm (almost 3 feet) in height and width, and makes an effective low hedge.

As with all volatile oil plants, lavender should be harvested in the early morning, to capture the oils the plant has replenished overnight.

MELISSA OFFICINALIS – LEMON BALM

Melissa means 'bee' or 'honey' in reference to its attraction to pollinators as well as its healing properties, just like honey. The herb is one of the mint family and has antiviral and antibacterial properties, and is also largely used for anxiety, depression and nervous tension. Its leaves, stems and flowers can be used in herbal infusions, creams, lotions, tinctures and massage oils to balance oxidative stress (see page 19).

This bushy perennial is very easy to grow, reaching up to 1 metre (more than 3 feet) in height in poor, well-drained, scrubby soils either in partial shade or full sun.

MENTHA PIPERITA – PEPPERMINT

Peppermint contains menthol, Omega-3 fatty acids, calcium, vitamins A and C and a selection of minerals. All aerial parts are steam distilled to make peppermint oil. Soothing and cooling, peppermint's medicinal properties help balance oil secretions.

The plant prefers fertile, moist soil in partial shade, and can grow to a height of 90cm (almost 3 feet) with underground runners that can spread invasively. Many gardeners grow mint in containers or planters rather than an open flower or vegetable bed for this reason.

OENOTHERA BIENNIS – EVENING PRIMROSE

All parts of the evening primrose, native to North America, are edible and used in medicinal preparations. The seed oil, being high in fatty acids, gamma-linolenic acid (GLA) and Omega-3, is naturally anti-inflammatory and skin rejuvenating. It is used in anti-ageing serums.

The plant naturalizes readily and grows wild in stony ground, or it can be sown in poor soil in full sun. It reaches 1 metre (more than 3 feet) in height and about 30cm (12in) in width. The sunshine-yellow flowers open in the evening and are attractive to moths as well as bees.

PAPAVER SOMNIFERUM – OPIUM OR BREADSEED POPPY

Poppy seeds are nature's gentlest exfoliators, due to their tiny size and high content of linoleic acid. In view of the detrimental impact plastic microbeads used in toiletries and skincare products have on marine life, poppy seeds offer a sustainable, biodegradable solution to exfoliation.

The plant provides attractive and colourful flowers in the borders and is very easy to cultivate from seed. Poppies often self-seed in disturbed ground at the edges of farmland or building sites, where tractors or diggers move the earth and the seeds germinate. Originating from the Middle and Far East, the plant thrives in poor soil with good drainage and can withstand droughts, bolting quickly from bud to bloom and seed pod in a few days.

Poppies are grown as agricultural crops on a large scale to provide seeds for the food industry, opium for the drugs industry and alkaloids for the pharmaceutical industry. The milky fluid that seeps from cuts in the green, fleshy and unripe poppy seed pods contains three alkaloids (morphine, codeine and papaverine) which, since ancient times, have been scraped off and air-dried to produce medicinal products.

The simplest way to harvest the stems and seed heads is at the end of summer, when they are brown and dry, on a wind-free day, and place them upside down in a paper bag to collect all the seeds. The seed heads also make attractive dried floral arrangements. Store the seeds in a sterilized jar for use in recipes (see pages 130 and 166).

PELARGONIUM GRAVEOLENS VAR. ROSEUM – ROSE-SCENTED GERANIUM

A native of South Africa, the rose-scented geranium is a large (up to 1.5 metres/5 feet tall) shrub that can be cultivated indoors in winter and outdoors during the warmer months, or in a sheltered, warm spot during either season. It prefers a sunny, well-drained position and root cuttings can be made to propagate the plant in spring.

Geranium oil is used extensively in aromatherapy and massage for its regenerative properties for the skin and as a stimulant for the adrenal glands, which produce essential hormones. The oil is obtained through a steam distillation of the edible flowers and leaves and is naturally antibacterial.

ROSA GALLICA OFFICINALIS – FRENCH OR SHRUB ROSE

The French or shrub rose is also known as the Apothecary's Rose – and with good reason: the petals, fruit (hips) and seeds have multiple benefits for skincare. Rose oil is extracted from the petals and can be used as a serum or to make a hydrosol, or distillate, of rose oil and distilled water – a tonic to cleanse and refresh the skin. Rosehip oil (unlike rose oil) is pressed from the fruit and seeds. The hips are laden with antioxidants and essential fatty acids, including linoleic and linolenic acids, both of which are anti-ageing and moisturizing because they help prevent moisture loss through the skin cell walls, thereby keeping hydration levels higher in the skin. In addition, the hips contain polyphenols and anthocyanins that have antiviral, antibacterial and antifungal properties and may help reduce inflammation.

Rosehip oil is also high in skin-nourishing vitamins: vitamin A or retinol, which encourages skin cell turnover; vitamin C, which aids in cell regeneration, boosting radiance; and vitamin E, an antioxidant known for its anti-inflammatory effects. Vitamin A is also made up of a nutritional compound called retinoid, which, with regular use, is known for its ability to reduce hyperpigmentation or age spots (see page 21) and other visible signs of ageing.

Apart from *Rosa gallica officinalis*, other candidates for the skincare garden that yield large rosehips are: *R. glauca, R. canina* (the dog rose), *R. paulii, R. davidii, R. forestiana, R. macrophylla, R. moyesii, R. rugosa alba, R. setipoda* and the Scotch rose, *R. spinosissima*, which has purple-black hips. Most of these faster-growing roses can obtain heights of 3–5 metres (up to 16 feet) and a spread of 3 metres (nearly 10 feet). They self-seed freely, are quite invasive (especially the dog rose) and prefer to grow in the shade of trees, hedgerows, sheds and walls.

For making hydrosols, I use the most scented roses, including *Rosa damascena, R. centifolia* and the old and English rose varieties from David Austin Roses, such as 'Gertrude Jekyll', 'Boscobel', 'Constance Spry' and 'Desdemona'.

SALVIA OFFICINALIS – SAGE

Sage is a shrubby Mediterranean perennial that has been widely used in herbal medicine and cookery for millennia. The oils derived from the leaves by distillation are astringent, antiseptic and stimulating as a hair tonic or rinse and help to control dandruff. The purple flowers can also be used to make a herbal tea which both cools the body and provides a calming, restorative means of hydration.

Sage prefers neutral to alkaline soils and full sun. It is normally grown from softwood cuttings.

SALVIA ROSMARINUS – ROSEMARY

Rosemary is another evergreen plant that originates from the hot, dry Mediterranean region. A healing, anti-inflammatory herb with powerful antimicrobial, antioxidant, stress-reducing and skin- and hair-conditioning properties, its leaves and flowers are used to make an infusion to help eliminate dandruff, or an essential oil. A mixture of rosemary and almond oil massaged into the skin helps relieve nervous tension and relax muscles.

A bushy, upright shrub, rosemary can grow up to 2 metres (more than 6 feet) in height and 1.5 metres (almost 5 feet) in width. As with other Mediterranean plants, it prefers a sunny position and well-drained soil.

SYMPHYTUM OFFICINALE – COMFREY

Like yarrow (see page 34), comfrey is a classic wound herb and its oil maceration and poultice (crushed leaves) can be used to massage skin and alleviate pain from sprains and bruises.

The drooping purple flower heads and the roots of comfrey contain the compound allantoin. This helps to moisturize and soothe dry irritated skin, promotes rapid skin cell growth, contributes to skin renewal, protects against bacteria and reduces inflammation.

The plant can grow to 1 metre (more than 3 feet) in height and is a prolific and invasive self-seeder, often found in shady hedgerows and moist field margins.

NOTE OF CAUTION

*Comfrey should only be applied externally, and not on broken skin, as it contains pyrrolizidine alkaloids that are thought to cause liver damage if taken internally.

THYMUS VULGARIS – COMMON THYME

All aerial parts of this familiar Mediterranean herb can be steam distilled for the antiseptic, astringent, antimicrobial and antibiotic properties of its essential oil. Thyme oil is widely used in aromatherapy – it can be used as a skin disinfectant in the case of bites or cuts and as a topical skin application that is anti-ageing and antioxidant.

Common thyme is a low-growing plant, no higher than 25cm (10in), and it prefers full sun and free-draining soil.

TRIFOLIUM PRATENSE – RED CLOVER

This is a valuable and fast-growing perennial for the skincare garden. As a green manure it fixes nitrogen from the air into the soil, suppresses weeds and improves soil structure.

The distinctive dark pink-red flower heads are rich in flavonoids, which have antioxidant properties. A herbal infusion applied topically is beneficial in the treatment of eczema, psoriasis and other skin conditions.

Red clover is an attractive annual which can be grown easily from seed, with very little management.

SKINCARE PLANTS TO FORAGE

Whenever I forage, I make sure I follow an expert's guide and the foraging code: to take only as many leaves, flowers or buds as I need for a specific formula; never to pull a plant out of the soil or the sea bed by its roots; to collect seaweed that is already washed up on the shore; to pick lower lying plant material, leaving plenty of higher berries, seeds, fruit or hips for birds and other wildlife. I keep to legal, local byways and footpaths, never venturing onto private land without permission from the landowner. It's important to forage away from main roads where there is pollution, and to wash all plant material until the water runs clear, to ensure it's not contaminated.

The countryside code is simple: be respectful of the land; do not trespass; leave nothing but your footprints; leave gates as you find them; do not disturb livestock; leave plenty of plant material for insects, birds and other wildlife to feed on, especially in winter.

You will need a basket, a sharp set of secateurs and gardening gloves, in case of thorns or thistles along the way. Normally, I forage early morning, before the sun is too high, so that the plants do not wilt readily after picking. If I am not going to use the plant material immediately, I place its stems in a tall jug of water, in a cool, shaded place.

You will find a simple trug or basket, garden snips, gloves and a cloth or cover useful tools for your foraging activities and do take a field guide. There are a number of handy wild flower and foraging books and apps I consult to ensure that every plant I collect is identified correctly.

SEAWEEDS

Since ancient times seaweeds have been used for their cleansing and therapeutic properties. In what is known as thalassotherapy, the ancient Greeks used the healing properties of seawater and algae (including seaweed) in cosmetic and health treatments.

A seaweed bath is beneficial because seaweeds are humectants, drawing moisture to the skin, and contain high levels of lipids, proteins and minerals (such as zinc, iron, magnesium, potassium, sodium and selenium) which have great benefits in healing and repairing the skin. Vitamins K, B, A, C and E help plump the skin, even out skin pigmentation and promote the production of collagen. The high iodine content of seaweeds is deeply cleansing and by using dried seaweed powder in a salt scrub it is also an effective exfoliator, removing dead skin cells so new ones can form.

All along the British coastline, you can forage very successfully for seaweeds at low tide, from April to late September, particularly on rocky shores. I find lengths of bladder wrack, and a huge number of other seaweeds, washed up along the Jurassic Coast, from Sidmouth in Devon to Chesil Beach in Dorset.

There are a huge number of seaweeds which can be harvested along the coasts of Britain – British and Irish seas are home to more than 600 species of seaweed; this is more than 6 per cent of the known species globally.

On just four beaches local to me – Burton Bradstock, Cogden Ringstead and Osmington –

apart from the ubiquitous bladder wrack, I have been able to forage for all of the following – each with fascinating common names: berry wart cress; dulse; forest kelp; horned wrack; red comb weed; red grape weed; rosy fan weed; sea beech; sea oak.

I take a large net bag to collect washed-up seaweeds, and when I get them home I wash them in copious amounts of fresh, running water until it runs completely clear of sand or grit. I then soak them overnight in more clean water and rinse again the following morning. Finally, I hang the seaweeds pegged outside, just like laundry, to dry in full sun. Once fully dried, I cut them into small pieces with a sharp pair of scissors and I store the seaweeds in a sterilized jar or covered box lined with parchment where they will last for 6–12 months.

Seaweed also makes an excellent organic fertilizer for your garden and can be spread over your flower or vegetable beds.

GALIUM APARINE – CLEAVERS

Cleavers are the common sticky weed known as bedstraw or goosegrass that is ubiquitous in countryside ditches, along footpaths, on abandoned plots or at the base of trees. It's an invasive weed that creeps over, up and around other plants with its soft, barbed stems.

The diuretic effect of cleavers is thought to relieve swelling and promote the movement of fluid throughout the body. The plant is rich in vitamin C and has astringent and anti-inflammatory properties, providing an effective treatment for psoriasis and inflammation of the skin. You can use both stems and leaves to make a refreshing, hydrating herbal infusion and drink it warm with honey and lemon or cold with ice, cucumber, mint and elderflower cordial.

LAMIUM ALBUM – WHITE DEAD NETTLE

So-called dead nettles can be foraged easily in spring and early summer, as they can be found growing in abundance along roadsides, hedgerows, fields, woodlands and gardens. (The leaves have no stings so you can gather them easily without harm.)

Rich in vitamins, minerals and Omega-3 fatty acids, a compress of white nettle leaf infusion acts as a natural astringent, helping to regulate natural sebum production and treat acne.

Used in creams, white nettle leaves accelerate the healing process of wounds and burns, helping with skin allergies and eczema.

MALUS SYLVESTRIS – CRAB APPLE

The progenitor of the cultivated apple, *Malus domestica*, it is well worth looking out for wild crab apples in autumn – they look just like miniature apples in hues of yellow, pink, red and amber. Wild apples are found mostly at the damp edges of forests, in farmland hedges or on very extreme, marginal sites or even in ancient orchards.

Bitter to taste and quite hard, the crab apple's antimicrobial, astringent and exfoliating properties are enhanced when it is made into crab apple cider vinegar (see page 132). Applied topically, the acetic acid in the resulting toner helps remove dead skin cells, leaving skin feeling clean and soft, and it is effective in the treatment of acne.

TARAXACUM OFFICINALE – DANDELION

Dandelions take their name from the French words for a lion's teeth – 'dents de lion' – and although they might be regarded by some as common garden weeds, they are, in fact, a very valuable resource. Found growing freely in lawns and uncultivated ground, on the margins of fields and all along hedgerows and pathways, dandelions are also cultivated as an early source of food for bees and also for salad leaves.

Rich in vitamins A, B, C and D, and in potassium and magnesium, the flowers and leaves can be juiced or infused and applied topically to help stimulate and clear the liver of toxins, by acting as a tonic and diuretic. This helps relieve acne, psoriasis and eczema. A decoction can be made by simmering finely chopped clean roots in water for 15 minutes, for the same purpose.

Dandelions self-seed prolifically, and the plants can reach 30cm (12in) in height in the wild, especially in fertile soils.

These are five of the main plants I forage for skincare at different times of year but of course there are many more. The next chapter takes you through the four seasons in the skincare garden.

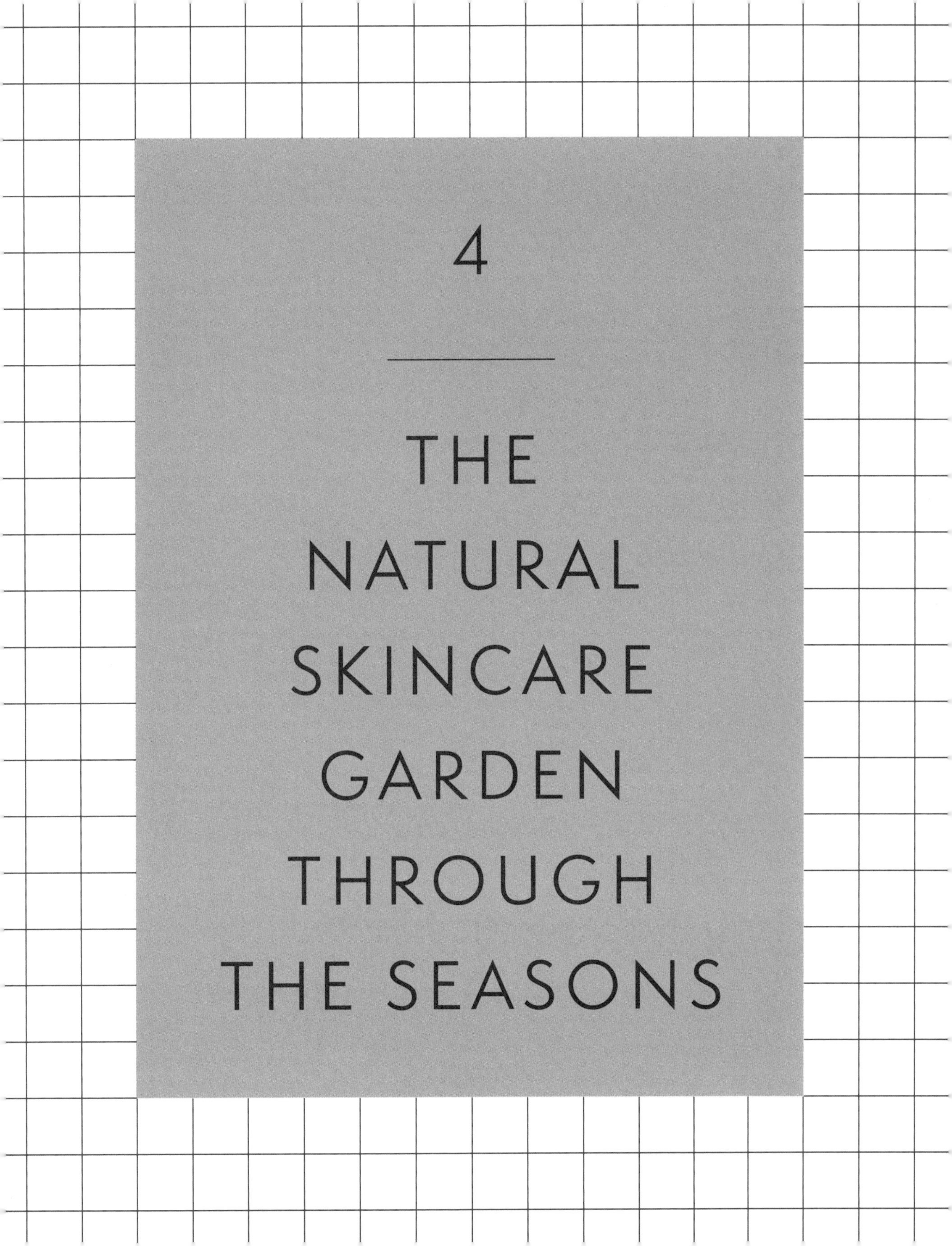

4

THE NATURAL SKINCARE GARDEN THROUGH THE SEASONS

WINTER

I begin this chapter with winter as, surprisingly, this is actually a beguilingly busy season for gardeners.

The great advantage of plotting and planning a new garden in winter is that there is plenty of time to sketch, list, measure and make decisions – invaluable activities with which to embrace the spring planting with purpose, vigour and focus.

THE PLAN

To plan my workshop and allotment skincare gardens, I focused first on those plants with the top botanical properties needed to create wonderful soaps, creams, balms, bath products, teas, hydrosols, macerations and serums. The active botanical properties needed for skincare products are anti-inflammatory, antibacterial, antioxidant, astringent and hydrating. In addition, plants like roses, lavender, rosemary and sage also provide fragrance through essential oils and hydrosols.

Above all, I feel it's very important that skincare manufacture has a sense of place and a seasonal focus. We have come to learn that eating local food is best, that seasonal fruit and vegetables taste better and that organic food is healthier for us – so why would we not have the same standards for skincare and toiletries?

We know that herbs have been grown for apothecary and skincare applications for thousands of years. In monasteries across Europe right up to the fifteenth century, early manuscripts, herbal and distillery books bear witness to the use of herbs to provide solutions to many skin ailments and to promote good health and mental well-being.

There are scores of plant varieties that can be cultivated in a skincare herb and flower garden, but it's always better to start with a core list of targeted plants than get swamped by unrealistic ambitions that don't come to fruition or go to waste. You should plant the amount of herbs that you can manage and that you need for creating the skincare products on your list – as you read through this book, you will gain a better and more accurate understanding of yields per plants and the quantities of each plant required to produce certain products.

Over and above those quantities you can also use plants to cook, make cordials, salads, flower arrangements, preserves, pot pourri, dried flower posies and so on – the applications are only limited by your imagination and time.

Different parts of the plant (roots, stem, leaves, petals, hips, berries) are used according to their properties and this is all taken into account during the planning. The final height and spread of the plant are also considered with regard to the length and width of planting rows, ease of access for harvest and distance from the source of water. (Of course, the more drought-resistant the plant the further from a water source it can be planted.)

Most herb rows in my allotment are 90–100cm (around 3 feet) wide and roughly the same distance apart – this is my optimal working distance. Your own skincare garden may need to be arranged quite differently, but space can usually be found alongside herbaceous borders, or in containers or hanging baskets on a patio, or in pots in a

conservatory or greenhouse, or on a balcony, window sill or roof top. There is virtually nowhere that is sunny, free-draining and sheltered that is not suitable for growing herbs. These adaptable and obliging plants require little maintenance, and many (such as scented pelargoniums, rosemary, lavender and sage) thrive on neglect.

THE COMPOST HEAP

All good planting starts with good soil, and good compost is vital to create sufficient soil coverage in the No Dig system (see page 27). A functioning, healthy soil has the ability to sequester carbon as well as grow nutrient-dense herbs and flowers, through microbial and fungal action. Compost inoculates soil with the organisms that build stable humus – the organic component of soil. Humus provides the single most efficient storage of water, minerals and carbon and it contains humic acid, which helps stimulate plant growth.

In summer, when air and soil temperatures are high, the decomposition rate in a garden compost heap or bins is fast. In winter, however, you may need to cover the compost heap or bins with a wooden lid, cardboard or thick newspaper, held down with stones, in order to retain the heat needed for microbial activity.

There are many schools of thought that relate to composting, but most agree that the optimal temperature of the centre of a compost heap should be 45-55°C (113–131°F), and at this temperature most pathogens will die.

If you are just starting a brand new compost heap at the bottom of the garden, in a corner of a courtyard, or just outside the back kitchen door, there are now so many solutions available online – from plastic containers, to metal wormeries, hot bins or wooden boxes, woven willow panels or wooden pallets that you assemble into squares. Or, if you have enough space to be able to hide a big 'compost cake' at the bottom of the garden, you can start a pile directly on grass or bare soil, marked out in a circle made out of bamboo canes or hazel sticks and garden twine, which help contain the pile in an upright position, like a mound.

The ideal base for your heap would be soil – so that worms, soil fungi and bacteria can help the decomposition process by digestion – the sign of a healthy and balanced compost heap is always in its vitality as a home of biodiversity.

Ideally, you should have an even mix of small pieces of both green and brown plant material in your compost heap, layered like a lasagne, that stay moist, aired and warm during decomposition. This could consist of alternating tiers of the following: grass mowings, dry leaves, clippings of twigs and

WHAT TO ADD TO COMPOST

- Manure from herbivores – horses, sheep, cows, goats, etc. – rotted or fresh
- All cardboard and untreated paper – newspaper, office paper and fliers or brochures that do not have a glossy finish
- Clean paper or fabric towels, cut into small pieces
- Coffee grounds and tea leaves
- Dust from the vacuum cleaner or sweepings
- Dried crushed eggshells (but not eggs)
- Fireplace wood ashes – but not coal ash
- Fruit and vegetables and their peelings from the kitchen or garden
- Grass lawnmower clippings
- Hay and straw
- Leaves
- Nut shells
- Seaweeds

WHAT NOT TO ADD TO COMPOST

- Meat, fish, egg or poultry scraps (raw or cooked)
- Dairy products
- Animal fats or any oils or pet faeces
- Citrus fruits
- Coal or charcoal ash
- Diseased plants
- Thick, fleshy leaves
- Perennial weeds or weed seed heads
- Plant material that has been sprayed with pesticides
- Plastic or dirty food takeaway containers

branches, garden prunings, fallen fruits, kitchen vegetable peelings, shredded newspaper, torn cardboard, ashes from a fire and dust from a vacuum cleaner.

For an average size domestic compost heap, covered with a lid or plastic tarpaulin, it's not really necessary to turn the compost heap or water it frequently. If you want to turn it with a garden fork, it will increase the aerobic decomposition, but generally a small heap, especially if placed in the sun, will decompose within 6–12 months. I leave the lid half on if rain is due, and then shut it again when the weather is dry. Balance is key: not too wet and not too dry. You can feel a good crumbly texture between your fingers when the compost is ready – it will be dark, smelling of earth and greenery, soft and lightly moist to the touch, with no slime or thick pieces of vegetation.

If you don't have access to a compost heap, and have no space for making your own, there are a number of good peat-free compost alternatives that can be purchased online, and which are listed on page 189. (Peat bogs are home to a rich biodiversity which should be protected and not disturbed. Once peat is dug out, it starts emitting carbon dioxide and methane (greenhouse gases) into the atmosphere.) As much as possible, I try to support my local plant nursery. The independent shop mantra is, 'If you don't use it, you lose it' and one outcome of the pandemic is that we have realized how important it is to spend money locally.

PROTECTION FROM PESTS

Most herbs and annual flowers are, by nature, quite resilient and hardy. They have few pests and diseases, require little in the way of maintenance, are fairly drought-resistant and some self-seed. Many have cut-and-come-again flowers – the more you harvest the flowers, the more the plant produces because the flowers are its reproductive organs, and vital to its life cycle.

If you are planning your herb garden next to your vegetables, it is good to know that many skincare plants are excellent companion plants. Sown alongside crop plants they help to deter pests and increase pollination of trees: for example, the scent of lavender and roses attracts many pollinating insects; mint, sage, rosemary and thyme deter cabbage moths, carrot flies and bean beetles; borage repels tomato worms; calendula attracts hoverflies and ladybirds that eat aphids.

By working with invertebrates, the ecosystem and soil structure of the skincare garden look after themselves: birds eat snails; frogs and hedgehogs eat slugs; spiders eat flies; snakes eat mice; worms aerate the soil, and so on. They are the workers in the field and the self-propelled biological pest-control battalions.

BUG HOTELS AND NESTING BOXES

Early winter is a good time to build a bug hotel. I use bricks, bamboo canes, old pots, garden stones, logs, twigs, pine cones, straw and slate tiles to create a safe, protected habitat for all kinds of invertebrates that need to over winter out of the rain and cold. A protected, south-facing wall is best and if you can find old slate or terracotta tiles to line the top it will prevent rain from penetrating into the structure. There are lots of small, ready-made bug hotels available online that can just be hung on a garden fence or fixed to the side of a garden shed. Old wooden pallets stacked one on top of each other can also make a useful structure, with the top laid as a green roof with nectar-rich plants to provide food once the bugs emerge.

The bug hotel serves as a useful residence for ladybirds, who are voracious feeders, eating up to 40–50 aphids every day. A single adult ladybird probably eats 5000 insects and insect larvae in its lifetime. Each female produces around 1000 eggs during a mating season, and the larvae feed on 400–500 aphids during the 2–3 weeks before pupating and coming out as young ladybirds.

Hedgehogs, newts, toads and song birds, such as thrushes and blackbirds love to eat slugs and snails, so they are welcome visitors in my garden. I attract birds with seeds, seed fat balls, water and nesting boxes.

I remove slugs and snails by hand, wearing gardening or plastic gloves – they are often found at the base of pots, on damp walls or under fleshy leaves of plants they love to eat. I then offer them to birds as the antipasti course.

THE TOOL KIT

If you don't already have a garden shed filled with tools and gardener's kit, you don't need to worry because the expense for setting up a skincare garden is small. My main tools consist of: gardening gloves, small fork, trowel and hand tools, hoe, rake, garden knife, wheelbarrow, watering can, bamboo canes, plant labels, secateurs, buckets, wicker flower baskets, seed box, twine and trug.

Many tools can be found for sale online, in reclamation yards, antique fairs, vintage shops, charity shops, car boot sales, junk yards and also on eBay or Facebook or local websites. Once you start delving there is no shortage of people looking to downsize from bigger properties to small flats or retirement homes where they no longer need their garden tools.

SPEAR & JACKSON
STAINLESS STEEL

THE SEED BOX

The cold months are also the best time for seed buying and organizing. I order my seed box alphabetically to keep track of what I have and cut down on waste – I reorder seeds every winter (see my list of seed suppliers in the Resources section on page 189) and ensure every variety is used before more is purchased.

Many hardy annuals self-seed with wild abandon – including clover, yarrow and comfrey – and the mint spreads profusely by root. The first year of planting is always a trial and error exercise – during the second year, you will have a better understanding of which seeds need to be bought in bigger quantities.

VERNALIZATION

Cold, frosty mornings are of benefit to the dormant winter garden. Frost breaks up clumps of soil into finer, more manageable tilth and some volunteer wildflower seeds require what is known as vernalization (the triggering of the plant's flowering process by exposure to a period of prolonged cold during winter) in order to germinate in spring. Overwintering aphids, slugs and snails perish in the cold snap and fungal problems are often cleansed by the fall in temperature. The garden is at rest and winter is an important time for all living creatures within it.

Herbaceous plants stop growing and die back, storing the carbohydrates they made from water and carbon dioxide during summer in their roots as starch and cellulose. This store provides the plants with a reserve of energy, designed to give them a head start in spring.

The onset of the coldest winter weather triggers enzymes in the roots to convert the starches back into soluble sugars. This means energy can be moved to the growing tips of the plant, ready for growth in early spring, enabling them to push their shoots towards the sunlight ahead of surrounding plants.

HERB DRYING AND STORAGE

During the late autumn and winter months, I will have herbs drying indoors. They need enough dry air circulating around them to prevent moulds forming. Bunches will be suspended on drying hooks in the kitchen, garage, barn or airing cupboard. The heat needs to be kept constant, and once all plant material is completely dry, I can start removing leaves and petals from the stems and storing the dried herbs in jars.

I sterilize 1–2 litre (1¾–3½ pint) jars with lids in hot water, drying them thoroughly. I cover the workshop table with newspaper and go through each herb bunch, tearing off dried leaves and flowers in small pieces. I fill the jars, then seal and label each with the botanical name of the plant, its common name and the date. I store the jars on open shelves and use the dried herbs for oils, salts, tonics and teas.

REFERENCES AND RESOURCES

The winter months are invaluable for reading and researching – I am habitually scrolling through herbal gardening and skincare blogs, books and magazines, reading up about the latest natural skincare finds, products and experts. There is a wealth of information about apothecary, medieval and monastic herb garden design and practice once you start searching online and in bookshops. You never finish learning, the plant-based world is full of innovation and inspiration – and the long winter nights are the perfect time to catch up on knowledge.

Some of my favourite botanical skincare producers and gardeners with commercial horticulture plots of all different sizes and types are listed in the Resources section at the back of the book – I follow them on social media and in blogs or subscribe to their newsletters so that I don't miss seasonal hints, tips or ideas.

SEED BUYING AND SOWING

By placing seed orders early in winter you are assured of getting the varieties you want – seed companies do run out quickly of popular plants. Again, at the end of the book, you'll find a list of the main companies that specialize in herb seeds in the UK and also in Europe.

I purchase my seeds from wholesale seed merchants and I calculate the quantity of plants needed to fill the number of metres per row in order to fill the patch entirely (that way there is very little space for weeds to flourish and compete for soil nutrients). In the workshop courtyard garden I sow and grow plants to form attractive groups, but on the allotment I sow in lines rather than clumps – this makes managing the plants and harvesting ergonomically quicker.

The less room for weeds, the less need for hoeing. Those parts of the skincare garden that are not sown with herbs will be covered with phacelia (see right), a member of the borage family, as a green manure that will help fix nitrogen into the soil. Nitrogen is essential for plant (and animal) growth, but it cannot be absorbed directly from the atmosphere. Instead it is converted by so-called cover crops that include legumes and clovers. These use microbes to 'fix' the nutrient in nodules on their roots so that it can be taken up from the soil by other plants.

CITRUS FRUIT IN WINTER

During those months when log fires are burning, the stove is always warm from cooking and the central heating is on, I take advantage of the heat and dryness indoors to dry bergamot, lime, lemon, orange and clementine slices. The thinly cut slices are pegged up to dry on a washing line indoors, placed on baking sheets in a warm oven (90°C/194°F) for 5 hours or on wooden racks with slats for the air to pass through. If you have an Aga or dehydrator that makes drying even easier.

Dried citrus slices can be used to fragrance plain sea salt for bath salts usage; finely grated peels can be used for face and body scrubs; fresh slices can be used for herbal teas, and fresh juices are used to make citrus facial tonic waters. Citrus peels can also be turned into tinctures by steeping in alcohol. At a time of year when the herb garden is at rest, the citrus bowl is a useful resource.

SPRING

INDOOR SOWING

Gardeners everywhere know that, despite the spring solstice of 20th March heralding the season of rebirth and growth, March in the northern hemisphere can be a cruel and cold month. Most of the UK is designated as Zone 9 on the USDA Plant Hardiness Zone map, the standard used by gardeners and growers alike, which means that we tend not to go beneath –6°C (21°F) in ground temperature at worst, even if it snows and ice covers the ground in early spring. It's the bitter winds, freezing rain and long darkness, however, that prevents most gardeners being able to make headway in their outdoor planting until at least the end of April or even May.

This is why I calculate backwards – if, in theory, the last of the frost risks in Dorset will be felt in mid-May, I calculate eight weeks before that, to mid-March, and I start sowing the seeds of half-hardy or tender plants indoors, as well as some of the hardy annuals, in order to get ahead with the planting schedule.

You don't need a polytunnel or greenhouse to sow seeds early – a sunny window sill, front porch or cold frame in the garden, a spot next to a south-facing wall or in a sheltered courtyard are all you need – as many annual seeds will germinate at around 18–20°C (64–68°F), which is where my central heating thermostat is set for most of winter.

I use either my own homemade compost or purchase from a good supplier (see the Resources on page 189). My favourites are organic, fine-textured, peat-free compost blends made from bracken and sheep's wool. (Wool naturally contains high levels of nitrogen, which acts as a slow-release fertilizer and aids water retention to give seeds the perfect start.) You can add horticultural grit to help drainage and root development, but for the first few weeks of life a general, good-quality compost will have all the nutrients and all the structure that a seedling needs.

Rather than plastic trays or containers, I use old cardboard egg boxes, coir pots and trays, toilet roll insides, cylinders made out of old newspaper sheets (which are fully biodegradable) or clean terracotta pots to sow my indoor seeds. I arrange them on watering trays on the front porch window ledges and fill almost to the top with compost, leaving a couple of centimetres for the top up coverage.

I sow seeds to a depth according to their size – very small seeds are barely covered with a dusting of compost, whereas bigger ones are pushed into a 1cm (½in) hole made by a dibber, or the end of a pencil, and then infilled with compost. For the first watering, I tend to water with a watering can with a fine rose to ensure they have a thorough soaking, but thereafter I tend to spray the surface of the compost with a plant mister.

The frequency and amount of watering depends on the temperature of the room, and whether there is constant central heating or not. As a rule of thumb, if you touch the compost with your hand and it feels dry, then spray it with water, but hold back if it's still moist from the previous watering. Overwatering seedlings is one of the most common causes of their death.

Calendula

In order to ensure a high germination rate and low wastage, seeds need to be kept at a temperature of around 18–20°C (64–68°F) in a light position, preferably south- or east-facing, so they obtain optimum light levels from morning to sunset.

Seedlings are ready to be planted out once they have around 4–5 leaves. At around the third week of April, I look closely at the barometer and weather reports and if the weather is clement and mild, I put out seedling trays to harden off outside, in a protected, sheltered spot where the night temperatures don't sink too low. About two or three weeks later, depending on the temperatures, they are ready to plant out into planters, the garden or allotment.

SPRING OUTDOOR SOWING

Ideally, the spring soil is prepared as soon as the weather allows – you need to start on a day that is frost free and not too damp, so that you are not walking on wet mud or sowing seeds in a mouldy grave.

My garden is run along permaculture lines, with a holistic approach to sustainability, biodiversity and resilience. No-dig gardening is practised, so that the soil structure is undisturbed and plenty of compost and leaf and bark mulches are created from garden waste and spread on the surface of the bed to a height of around 5–10cm (2–4in) annually, in early spring. (See chapter 2, page 27)

The seeds that were sown indoors can go straight into the soil along with their biodegradable containers, which decompose naturally once covered with compost. This saves having to handle the young seedlings. Hardy annual seeds are also sown in situ, in succession, in the ground, to a depth twice the size of the seed, or around 1cm (½in).

If you want botanicals all summer long, it's a good idea to carry out succession sowing every 2–3 weeks, so that there is a sequence of more seedlings ready for hardening off and planting out. Generally, in an average year, I sow three relays per spring, each three weeks apart as follows: indoor in early to mid-March (planting out beginning of May), outdoors in mid-April and again outdoors in mid-May. The harvests will then occur, approximately between June to September, continuously, as the plants are all cut-and-come-again: the more flowers you cut, the more the plant will produce.

The seeds are sown outdoors carefully and thinly, 5–10cm (2–4in) apart, as this will save me time not having to thin them out. I then rake the soil to cover the drill and firm using the back of the rake. I drench each row with a generous puddle of water, but I want to encourage the roots to grow deep into the soil structure, looking for water, so I only water thereafter when the soil is really dry, gauging the need. If you pamper plants with too much attention, the roots remain on the surface and in a period of drought that could mean the end.

Rows are staked with hazel and bamboo canes and string because as the plants grow you want as little of the material to touch the ground and get dirty. You need to distil plants that are as clean and undamaged as possible, so giving them a structure on which to grow upwards is crucial.

When creating the layout of both of my gardens I planned metre-wide rows of alternating plants, with pathways wide enough to enable hoeing and seasonal cut-and-come-again harvesting and succession planting. The pathways can be mulched with chip bark to suppress weeds.

No chemicals are used in my garden – by creating biodiversity and maintaining the health of the soil with compost, manure and wood bark mulches, plants grow strongly with natural resistance to disease and pests.

MAKING SEAWEED, NETTLE AND COMFREY TEA FERTILIZER

Living so near to the Jurassic Coast, I am able to harvest a wide range of seaweeds all along the coast. Around my village there are also fields rich in comfrey and nettles I can forage. Every spring I make an infusion by half filling a drum, bin, bucket or barrel with these foraged ingredients, using stems and leaves and rinsed seaweed (in case there is any sea salt left). Together they create a valuable source of the trace elements iodine, copper, manganese, potassium, phosphorus, iron and zinc. I top up the container with rainwater, put the lid on and leave aerobic and anaerobic bacterial decomposition to create a rich (albeit smelly!) plant fertilizer to pour over established plants after watering. I check on it every week or so, lifting the lid and giving the contents a good mix with a rod or stick.

This fertilizer tea takes around 3–4 weeks to make, depending on the temperature, and should be diluted 1 part tea to 10 parts water when feeding established plants. I feed once a month in the growing season, in the evening or early morning, after giving the plants a good overall soak with water. The tea is applied directly to the soil and not all over the plants, because the foliage and flowers will be used to macerate oils and so I want them clean.

The fertilizer can be decanted and stored in a clean jerry can or plastic container, in a cool, dark place. It should all be used within 6 months.

CUT, DRY, STORE, REPEAT

It's important to keep up with the flower and leaf harvest so that there is a succession of plant material throughout the growing season.

Cutting and cuttings

Late spring is the perfect time to start the cutting and propagation of the skincare plants in my plot. The plants are now coming to a crescendo of development and there is enough plant material to supply the workshop bench and for increasing stocks.

Picking

Remember: the more flowers you pick, the more the plant will produce them.

Try to pick first thing on a clear, dry spring morning, when the plant material is fresh and cool, with as little moisture as possible, so that they will keep better and not grow mould. Herbs produce their volatile oils at night, so the earlier in the morning, the better. At the peak of its season, the plant will have the highest concentration of its volatile constituents.

I cut a full stem, 10–20cm (4–8in), depending on the height of the plant. If I am using the material fresh, I place its stems immediately in cold water, and place the container in a cool, shady spot. Or, if I am drying the plants, I place them on a sheet of newspaper on my table to dry them flat or tie a string loop round the bottom of the stem and hang it upside down. (I go into more detail on successful herb drying in the Summer section.)

Propagation by cuttings

Taking cuttings is one of the simplest and most frugal ways of increasing the number of fleshier, softwood skincare plants, such as pelargoniums, lavender, rosemary and sage.

Following these easy steps, I get lots more free plants for my garden.

Collect non-flower shoots early in the day, before the heat of the sun dries the moisture in the plant.

Remove up to 10cm (4in) of shoot, cutting just below a leaf node or joint, above a bud on the parent plant. Here there is a concentration of nodal plant hormones to stimulate new root production.

Place the cutting material in a clean plastic bag with a name label, and store it in the fridge if you cannot prepare the cuttings immediately.

Using a sharp knife or scissors, trim below a node to make a cutting about 5–8cm (2–3¼in) long.

Remove all the lower leaves and pinch out the soft tip leaves at the top. You can dip the cutting end into hormone rooting powder, but that's not obligatory.

Poke a hole for the cutting in a container of compost and grit mix using a pencil and gently place the base of the cutting with the first pair of leaves just above the level of the compost.

Label the pot or tray and water it from above with a sprinkler hose.

Place the pot on a light, warm window sill with a constant temperature of at least 18°C (64°F), and water spray mist the new cuttings two or three times a week to keep the soil moist. You might cover with a propagation lid or plastic, but this isn't necessary. The main thing is to avoid really hot, direct sunshine or extremes of cold air.

Once the cuttings have rooted, let them get used to outside temperatures for a couple of weeks before repotting them on into their new containers.

Keep new cuttings watered and out of bright sunlight until they are established and strong.

Propagation by division

Perennial plants that grow in clumps can all be very easily divided by root division. I propagate all my large lemon balm, yarrow, rosemary, mint, thyme and sage plants in this way.

Early or late in the day, not in full sun, dig up the whole plant root ball with a garden fork.

Break the clump with your fingers, to divide the root ball into smaller sections. You may need a sharp knife to do so, or if the clump is too big, use two garden forks, back to back, as levers that separate the clump into two halves.

Break the original clump into as many smaller clumps as possible and then replant them separately, well spaced out, into planting holes to the same depth, using fresh compost. Cut back any bruised or damaged foliage and water well.

SUMMER

The height of summer is a very busy time in the skincare garden. All the herbs and flowers are in full growth and bloom and there is an intensity in the fragrances, colours, textures and movement of the garden. It's important to enjoy the fruits of your labours – gratitude and reflection are important elements on our well-being journeys. Breathe the floral scent of summer slowly and appreciate its aromatic range.

Tall plants, like borage, sway their star-shaped flower heads in the wind. Calendulas burst open into bright orange plates surrounded by a sea of lime-green, fleshy leaves. The sweet, evocative scents of roses, chamomile and lavender fill the evening air after a warm, dry day. Early morning cuttings of herbs like mint and sage feel like cool, soft velvet covered in dew on the fingertips. Rosemary and thyme are at their most potent, their aromatic leaves and flowers leaving a heady scent reminiscent of summer holidays in the workshop. Every day I take time to enjoy the bounty and abundance of botanical riches that this season affords me, before I set to work.

As summer unfolds at speed it's important to keep up with what I call the four Ws of gardening: watering, weeding, waiting and watching.

WATERING

Every year and every region is different, and, in Britain, the only predictable element about the weather is its unpredictability. If rainfall has been sparse, it's important to remember that most of the plants in this book are quite adept at dealing with drought stress. Climate change is playing a very important role in turning seasons almost upside down. Longer, drier, more extreme summer temperatures are now experienced by many of us, which, quite suddenly, are broken by thunderstorms, flash flooding and unseasonal floods.

Our ancient plant varieties have built in resilient ways of coping with stresses. For example, the waxy needles of rosemary and the hairy leaves of borage help to reduce water loss through evaporation. The silver leaves of lavender and the grey-green leaves of sage help to reflect the heat of the sun. The deep and invasive root system of the mint plant is adept at seeking out water deep in the soil. Clover's ability to suppress all weeds in its vicinity enables it to eliminate competition for water. The etiolated (spindly) leaves of St John's Wort and chamomile ensure little water is lost into the atmosphere through transpiration.

The no-dig gardening system, layers of compost and mulch and regular weeding ensure that summer water loss is minimized in my garden. To ensure that I minimize stock loss, however, I do water plants if I see that the soil is particularly dry over long periods. I water with a watering can late in the evening, after sunset, or early in the morning, and I focus the pouring right at the roots of plants or in containers, making sure that terracotta pots are kept moist (as they do tend to dry out most quickly). Little and often is best, but I never overwater, as this causes soil nutrients to leach away – and water is a very valuable resource too. Plants need to be encouraged to send their roots deep into the soil to search for water and over pampering them will only lead to their demise. Look after the soil in the first instance, and it will ensure the long-term resilience and longevity of perennial plants.

Annuals, on the other hand, are destined to complete their life cycle in one year, and long periods of drought may well result in them being unable to deal with the cellular stress, thereby bolting, or going straight from flower to seed in an accelerated time frame. High air and ground temperatures and lack of water induce the plant to redirect resources away from producing leaves, buds and flowers to prematurely producing seeds to ensure its reproduction.

With annual plants such as poppies, calendula and borage, I find there is a balance in the watering: to carry some of the plants forward in order to reap the benefits of a longer flower harvest, but at the same time to leave some of the plants to their own bolting devices, as I need to collect and dry seeds for the stock cupboard, for products and for next year's sowing. From year to year, the annual garden plants produce their own seed stock and no extra stock is bought in.

There is as much beauty in the seed pods and in the late summer evolution of the plant as there is in spring, so I take time to appreciate that as well.

WEEDING

Weeds do not stand much of a chance when you follow the no-dig gardening system and plant as intensively as possible. In the allotment, I use a long Dutch hoe and also a Niwaki Hori Hori knife and hand-held hoe to eliminate invasive perennial weeds, such as bindweed, thistles, brambles, oxalis and ground elder, as these are extremely dominant and can take hold of an entire patch if ignored, smothering small seedlings in their race for land grab.

Many of these perennial weeds are seemingly indestructible, and attempts to remove them completely will fail because of their ability, via roots, rhizomes or nodules, to disperse. Pulling, digging or hoeing may well help to redouble the infestation rather than halt it. Far better to suppress all light using black biodegradable film so that these weeds are unable to photosynthesize. But I leave many wild flowers and white nettles undisturbed, as they are important to wildlife and can be used in poultices, tinctures, macerations and teas too.

The old adage goes that there is no such thing as a weed, only a plant in the wrong place. So many plants that you might have been led to think are weeds that should be eradicated from the borders, are in fact sought after or even cultivated and used in herbal medicine as healing, calming or preventative toners, balms and medicines.

The skincare garden is all about balance and nourishing the soil and its natural biodiversity so that all manner of plants are able to grow and thrive in it – it's not all about obsessive tidiness and exclusion. Here are five common garden or hedgerow 'weeds' that are actually beneficial to the skin:

- **Plantago major – Plantain** The broadleaf plantain has long been used in traditional medicine. The compounds it contains have a cooling and soothing effect, making it useful for treating eczema, burns and scalds. The leaves can also be applied topically to wounds, insect bites and eye inflammations.
- **Stellaria media – Chickweed herb** The leaves and juice of chickweed can be applied directly to the skin to help heal irritations, itching, rashes and scabs. It contains demulcents, which provide a protective film, while saponins ease the itch. It can also be added to a warm bath to help ease the pain of rheumatism.
- **Rumex crispus – Yellow dock** The roots can be turned into a tincture and taken as part of a cleansing routine for irritant skin, rashes, eczema, acne and shingles.
- **Equisetum arvense – Horsetail** These weeds can be taken in decoctions, juices, baths, poultices and used as a mouthwash or gargle for their healing properties as wound herbs.
- **Arctium lappa – Burdock** Both roots and leaves are used in decoctions, poultices and infusions to create a facial wash for acne and fungal skin infections.

WAITING AND WATCHING

Patience is not my forte, and my eagerness to get the botanical crop grown, harvested and dried can sometimes overshadow common sense. You have to harvest at the optimal time, and that takes years of practice.

Most antioxidant phytochemicals (the chemicals found in plants), are polyphenols and flavonoids, or micronutrients, whose compounds and properties we want to extract from the aerial parts of the plant, either through distillation, infusion or maceration. Scientific evidence has shown that when a plant is in its reproductive state (in summer) its antioxidant properties are at their most powerful.

For collecting herbs and flowers that I intend to dry, I wait for the plant to be in full foliage flush and bud bloom and then harvest in either the early hours of the morning or later in the evening. On a warm day, if you brush your hand along rosemary needles, across sage leaves or over a lavender spike, you will be able to smell the unmistakeable aromas of plant oils attracting pollinators to their flower heads. Brightly coloured, newly opened roses, calendula, borage and clover are similarly at their peak during the heady summer days. You want to harvest when the bud or flower is just opened, not when it has been open for several days and is in decline.

Waiting and watching for the right moment is crucial in harnessing the full benefits of the plant – a few days too long and the energy of the plant might be drawn towards seed production.

DRYING

For skincare teas and infusions, I also tend to pick a few flowers in closed bud and a few very young tip leaves as these are sweeter.

To avoid bruising or damaging any of the plants, I collect them in large baskets or foraging bags and immediately take them to my workshop for drying. My first task is to check that all the plant material is clean and not carrying any beneficial insects, since I garden without pesticides.

The best way to remove any unwanted visitors is to shake stems outside or leave the harvest laid out flat, airing on a table or work bench. Small insects will crawl out of the foliage or flowers and you can then scoop them up gently with a dust pan and brush and put them back in the garden.

For soft-stemmed plants such as cleavers, dandelions and comfrey, you can lay leaves, flowers and buds to dry on a worktop that is covered with clean paper or linen tea towels. There are also professional, digitally timed dehydrators with stacking trays, with temperature ranges between 35°C–70°C (95–158°F). These can be used for larger harvests.

If you don't have a great deal of spare table or worktop space, the best way to dry herbs that have stronger stems, such as lavender, rosemary and yarrow, is by using clothes airers, hooks suspended from the ceiling, a Sheila Maid, herb drying towers or the inside of an airing cupboard.

Tie your plant material in small bunches with string or twine and hang them a good 10–15cm (4–6in) apart, to ensure air flow all around the bunch. You may want to place a clean sheet or newspapers underneath, to catch any needles, petals or fine seeds that may fall as the plants are drying.

THE ROLE OF TEMPERATURE, LIGHT, AIR AND TIME

If you try to dry your plant material too quickly, in intense warmth or sunlight, it will deteriorate rapidly and aromatic constituents will be lost as well as colour and quality. Drying is a skill – it requires balance in temperature, light, air and time. Herbs and flowers should be dried in darkness, to retain colour, with ambient, warm airflow between bunches, with as low air humidity as possible.

Herbs and flowers that may otherwise seem completely dry when you rub them between your fingers, are able to absorb moisture in the air and attract mould. Dandelion is particularly difficult to dry, as its leaves attract moisture from the air, even after drying.

It is experience and repetition that enables you to learn and to separate the fleshy leaves that need longer to dry (for example, sage), the biggest flowers that need to be broken up before laying down for drying (roses and calendula), the whole stems that need to be hung in smaller bunches (rosemary, chamomile, clover, borage and yarrow), and those that can be hung in thicker bundles (thyme and poppy, the latter with the seed pods down encased in paper bags to collect the tiny seeds).

Drying times for herbs and flowers take anywhere between 14 and 21 days, longer if there are humid spells. I keep turning flowers and leaves if they are laid out flat to ensure that both sides receive airflow. I am fortunate in my workshop to have a number of old Victorian wooden laundry drying racks – they can be found in vintage, antique and junk shops. Modern metal versions are also available from hardware stores.

Again, patience is of the essence since drying cannot be rushed – the waiting and watching advice rolls on throughout the autumn. As the plants produce more leaves and flowers I keep cutting in succession, because the successional sowings that were done in spring keep producing plant material in monthly waves.

Over the past decade, climate change has altered many seasonal patterns and in mid-autumn, if we are having a particularly sunny 'Indian summer', with long warm days and little rainfall, I take advantage and set up my trestle table outside to dry my plants.

AUTUMN

During autumn, you will find me busy harvesting and drying any remaining plant material for my workshop and preparing the garden for winter.

Because of climate change, the seasons no longer arrive neatly and on time – summer days sometimes seem to go beyond the September equinox of the 21st or 22nd, and we benefit from a long, warm and productive start to autumn, where the garden is filled with late-flowering calendulas, borage and chamomile (thanks to successional seed sowing), rosehips, apples and crab apples.

My to-do list looks something like this:

COLLECT SEEDS

Seed collection is a sure way of saving money: as soon as the seed heads are ripe, but before they have shed their seeds, on a warm and wind-free day, you will find me out in the garden with my trug and secateurs.

I pick borage, calendula, poppy and comfrey seed heads individually and lay them out to dry on paper-lined trays. This enables seed to be more easily extracted from the seed heads. I clean off the material surrounding the seeds (called 'chaff', whence the expression 'to separate the wheat from the chaff') as this could potentially rot.

STORE SEEDS

I place dry seeds in labelled paper packets or envelopes in an airtight container (like a glass jar or plastic lunch box with lid) and ensure that the seeds are not exposed to humidity or warmth because they can deteriorate or die from fungal disease or rot.

Seeds can be stored in a fridge at 5°C (41°F) for several years if they are not going to be used the following spring.

If the garden keeps providing, I keep cutting and drying (see page 95).

STAKE AND PROTECT

The coming winter winds and cold temperatures may wreak havoc in your garden, and there is nothing worse than losing plant stocks that you have looked after assiduously during the warmer months, only for them to be snapped off or die in late autumn or winter.

Make sure that you firmly stake larger plants with hazel or bamboo stakes and string. If you don't have an indoor sunny space for tender plants, such as scented pelargonium or lavender, you can fleece them – I use 35gms polypropylene fleece jackets which can be put over plants and secured with drawstring. These allow light and moisture through but protect tender plants from frost.

Some gardeners use straw as overwintering protection, which is an eco-friendly option but it can provide a safe haven for slugs and snails so you need to be vigilant.

START CLEARING THE GARDEN

In order to look after the wildlife in my garden I strike a balance between tidiness and conservation. The beneficial insects and predators in my garden need shelter during the cold months ahead –

ladybirds nest in the centre of my bamboo canes; masonry bees hibernate in holes in trees or in the ground; toads take shelter in log piles; butterfly eggs can be found in dried out grass stalks; hedgehogs spend the winter sleeping under leaf or twig piles. I leave out fat balls, seeds, nuts and shallow water trays for resident birds, making sure they are not accessible to vermin.

My pruning technique is to remove all dead plant material, as this will only rot and potentially attract disease, but to keep attractive seed heads or foliage that still has potential for beauty or wildlife shelter.

CUTTINGS AND DIVISION

Now is a good time to look at the garden carefully and assess the planting for next year.

Every year I take soft cuttings of plants that I need in greater numbers to fulfil more orders for skincare products, and I line them up on a sunny window sill or the south-facing potting shed where I overwinter them. I don't water too much, especially the Mediterranean plants, and I ensure that the ambient temperature remains steady and falls no lower than 15°C (59°F).

I divide the roots of mint, rosemary, sage and lavender plants, using peat-free compost and plenty of horticultural grit or vermiculite for drainage. It's a good opportunity to reuse all the plastic pots I accumulate during the year.

KEY

1. POPPY
2. CALENDULA
3. CHAMOMILE
4. CLOVER
5. YARROW
6. BORAGE

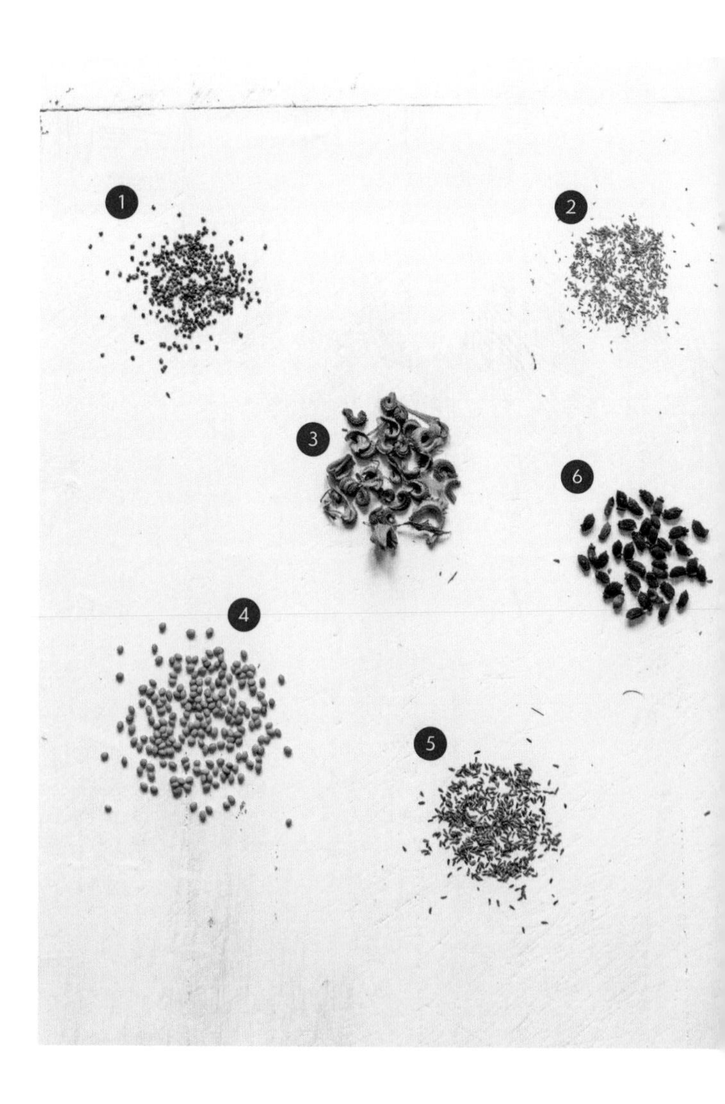

MAKE LEAF MOULD

My obsession with leaf mould knows no bounds, it is quite simply one of the very best soil conditioners (and it's free). Fungal decomposition of deciduous leaves (especially oak, hornbeam and beech leaves) creates a wonderful mulch for the autumn borders and your skincare garden will greatly benefit from this humus top coat filled with microorganisms.

I rake up as many leaves as I can from the allotment and around my village, put them in my wheelbarrow and then fill large hessian sacks (you can use bin liners) in which I have cut a few holes – moisture and oxygen are vital for decomposition. You could also create a leaf mould net with wooden stakes and chicken wire – just to hold them all in. The following year, in late spring, you can shovel the leaf mould onto your beds, especially around feed-hungry plants, such as roses and trees.

CLEAN AND CLEAR THE POTTING SHED OR GREENHOUSE

The autumn is a good time to Marie Kondo your outbuildings, if you are fortunate enough to have a potting shed or greenhouse. Here's what you need to do:

- Take all tools, pots and equipment out and be ruthless with decluttering.
- To maximize winter light, give all the glass panes a good wash with hot water, vinegar and a few drops of eco washing-up liquid.
- Brush away all soil, dirt and cobwebs, and reconfigure your garden building to optimize the working space.
- Use wooden crates, cardboard boxes and galvanised steel buckets as containers for tidiness and small hooks to enable you to keep your tools in order.
- Wipe down every tool with a stiff brush, hot water and soap and dry them thoroughly using a clean rag (recycled from fabric remnants). Sharpen the blades using a sharpening stone, apply some wood oil and metal oil to the handles and blades and give them a final rub with a soft microcloth.
- Terracotta pots also need to be thoroughly washed (I have been known to put mine in the hottest setting in the dishwasher!), in order to avoid the transfer of disease.

Ensure that all seeds and plants are safe from vermin or pests – check for any openings into your shed or greenhouse – and remove plant material that has black fly or aphids.

My potting shed also serves as a useful space for growing winter cut-and-come-again salad leaves and herbs, such as coriander, parsley, basil, fennel and dill, which are vital for skincare nutrients.

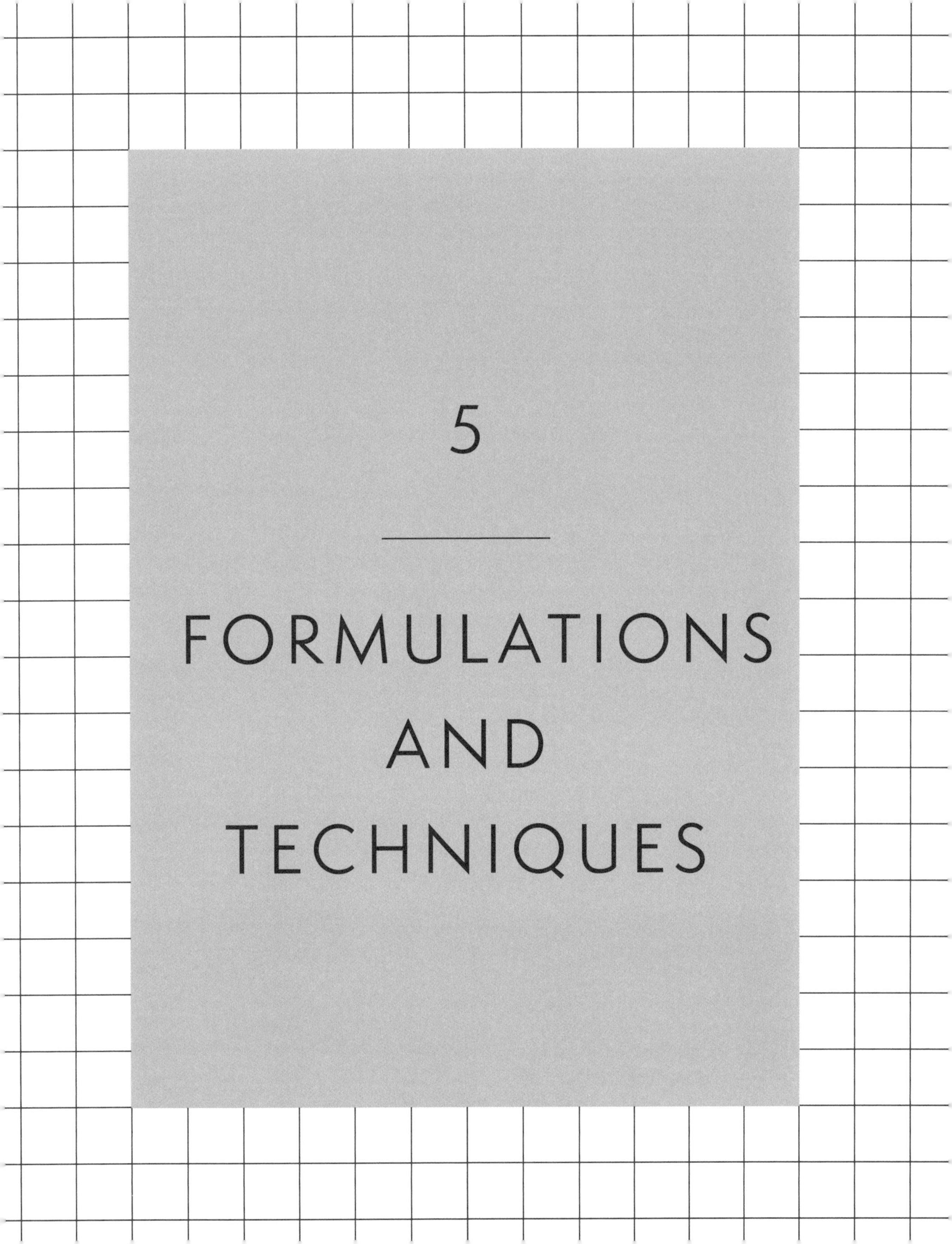

5

FORMULATIONS AND TECHNIQUES

SHELF LIFE

> **VEGAN WAXES**
> In handmade skin creams, formulators often use beeswax pellets, so if you are looking to create a completely vegan skincare product, you may wish to opt for candelilla, rice bran or sunflower waxes instead.
>
> For handmade creams, an emulsifying wax is used to emulsify the oil and water parts of the formulation into a homogenous cream. The one that is frequently used is a safe, vegan, food-grade product, commonly labelled as Steareth-21. It is a saturated fatty acid, derived from stearic acid, a naturally occurring acid found in both plant oils and animal fats. Another common emulsifying wax is cetearyl alcohol, a flaky, waxy, white solid that is made from a mix of cetyl and stearyl alcohols, which occur naturally in plants such as coconut palms, maize plants or soy plants.

Once you have mastered the basics of making your own infusions, macerations, soaps, hydrosols, enfleurages, tinctures and vinegars (specifically apple and crab apple cider vinegar, or ACV), they will prove to be really useful techniques that can be used with a wide range of different plants – there is no end to the different products that you can make for your skin. They really are simple, seasonal and effective.

The keys to homemade, small-batch skincare products for personal use are that they are made in little quantities, using fresh, seasonal ingredients, stored in sterilized containers and used in a short time frame. This limits the risk of bacterial contamination, as obviously fresh products need to be safe to use.

You can use natural preservatives, like tocopherol Vitamin E, if you want to keep any aqueous (water-based) products, such as infusions or hydrosols, for longer.

When you purchase skincare products, you will notice a little jar sign, or bottle sign, indicating the use-by date once the seal is broken. This is normally 12–24 months, as many lotions, creams and bath toiletries contain chemical preservatives and stabilizers that allow the product to sit on a supermarket or chemist's shelf for long periods.

Toiletries are normally kept in bathrooms, where warm, humid conditions could easily create the perfect breeding grounds for bacteria, moulds and yeasts.

With homemade products, it's really important that they are made safely and cleanly and used quickly, because every time air enters the jar or bottle, the likelihood of contamination rises. By ensuring that all tools, utensils and containers are completely sterile and that the products most likely to degenerate quickly are kept in the fridge (at a temperature of around 5°C/41°F), you will have a safe and efficient skincare product at your disposal.

GLASS COLOUR

To lengthen shelf life, you can use amber, blue or green glass, as these colours prohibit light from spoiling the ingredients and consequently the products last longer. Amber glass is the preferred colour of pharmaceutical companies and producers of essential oils and plant-based products. Cobalt-blue glass offers medium protection from UV radiation, and like amber, it absorbs UV radiation. However, it allows blue light through. Cobalt-blue bottles and jars have long been used for packaging apothecary and medicinal products.

Green glass blocks some UV rays, but is not nearly as effective as amber or cobalt-blue glass. Use green glass for products that are mildly sensitive to light, such as botanical oils.

KEEPING EQUIPMENT CLEAN

Skincare products need to be made in a clean and safe environment, so do ensure that your work space and all equipment is cleaned thoroughly after use and stored hygienically.

If you are making hair and body products that are kept in glass jars or bottles, sterilize the containers in the hottest setting on your dishwasher. Alternatively, place clean glass jars in an oven preheated to 150°C (300°F/gas mark 2) for 20 minutes, remove and cool. Sterilize lids by placing them in a heatproof bowl and immersing them in kettle-hot water. Dry using a clean tea towel.

WHAT YOU NEED

This guide has both frugality and practicality at its heart and I am going to share simple and effective tips to get you making your own botanical skincare products. Before that, there are a couple of steps to prepare your space.

First, the best start you can make is to be rigorous in assessing the present contents of your bathroom cabinet and dressing table and have a good declutter. Put to one side old products, those possibly past their sell-by dates, which by now may be contaminated with bacteria, or any that you no longer use, especially products where the ingredients read like a list of chemical names (more about them later!).

Second, give your bathroom shelves and dressing table a thorough clean. I've included in the section on Sustainability (see page 182) the best ways to clean with homemade eco-friendly products, so once you have made your own spray, you can deep-clean each drawer and shelf using a linen cloth. Then you can line them with paper, or insert little storage containers to keep everything in place. In the Resources (see page 189), I list a useful number of suppliers I use for bathroom items that are both attractive and practical.

It's definitely worth preparing a calm, organized space in which to keep your skincare products as it will provide pleasure and purpose to your daily routine.

A NOTE ON WEIGHING AND MEASURING

Professional skincare formulators use digital scales to weigh everything, including water and other liquids. As this book is intended for making skincare products at home, I have used recipe-style weights and measures to make life simpler; not everyone wants to measure all ingredients down to the last gram or millilitre.

As an approximation for dry ingredients:
1 teaspoon = 4–5g (⅛oz)
1 tablespoon = 14–15g (½oz)

A NOTE ON WATER

In all the formulations on the following pages, especially in soap-making, it is really important to use only distilled or de-ionized water if water is listed in the ingredients. Tap water can be filled with many impurities, that include: metals and minerals leached from the soil, such as aluminium, arsenic, calcium, nitrates, phosphate and sodium; copper or lead leached from pipework; chlorine, which is added to water as a disinfectant, or fluoride, which some local authorities add for dental health; or the element radium, which is released when rocks that contain radium break down. It's fine to use tap water for teas or for products that are used in the bath, but for all other preparations, opt for cosmetic grade distilled or de-ionized water, which can be purchased fairly cheaply online.

ESSENTIAL EQUIPMENT

Most of these items are probably in your kitchen cupboard already. If you are going to make a lot of skincare products – and you can afford to and have the space – keep a box of tools especially for your preparations, separate from daily kitchen usage.

- Chopping board
- Eye protection
- Electric blender
- Jugs with a good lip for ease of pouring
- Metal spoon
- Muslin cloth
- Pipette
- Pots and bowls, various sizes
- Rubber gloves
- Silicone baking moulds
- Tea pot
- Tea towels
- Thermometer
- Scales, preferably digital
- Screw-top jars – I recommend Kilner or Le Parfait
- Sieve
- Spatulas
- Whisk

INFUSIONS

When you infuse botanical material you are harnessing the properties of the stems, leaves, berries, bark or floral parts of a plant. Typically, this is done with hot water from the kettle and the ratio of plant material to water is normally 1 heaped teaspoon of dried material or 2 heaped teaspoons of chopped fresh material for every 200ml (7fl oz) of hot water. An infusion can be made and applied topically, or it can be drunk as a tea.

If you are making an infusion to drink as a herbal tea, heat freshly drawn water in the kettle to near boiling point, just below 100°C (212°F). Spoon the dried or fresh herbs into a warm teapot and pour the hot water over. Put the lid on the pot and leave to steep for 6–8 minutes, then strain the infusion into a cup. Drink with a slice of lemon or leave to cool completely and drink with ice – herbal tea is refreshing across the seasons, warm or cold.

BOTANICAL OIL MACERATIONS

INGREDIENTS FOR A 1-LITRE (1¾-PINT) JAR

300g (10½oz) dried herbs, flowers or hips (such as lavender, calendula, rosehips, thyme), finely torn or cut

500ml (18fl oz) vegetable oil (rapeseed, olive, sunflower, almond, hemp, borage or flaxseed)

Sterilize a 1-litre (1¾-pint) jar and lid in hot water and dry thoroughly.

Place the herbs, flowers or hips in the jar and top with the oil. Close the lid tightly and shake the jar a few times. Leave on a sunny window sill for 4 weeks or in a dark cupboard for 6 weeks.

Shake the jar regularly. When you want to use it, strain the oil through a sieve into a clean/sterilized jar or bottle and discard the herbs, flowers or hips. Seal, label and date. Oil macerations keep for up to 12 months in a sealed, sterilized jar/bottle in a cool, dark cupboard or bathroom cabinet.

BOTANICAL COLD-PROCESS SOAPS

Botanical soaps are created by the cold-process method, which means that very little external heat is applied – only enough gentle heat to melt the oils which, at room temperature in a temperate climate, are solid. The advantage of this method is that the curing process is slow and gradual, creating a more natural, mild and creamy soap, filled with moisturizing glycerine that is better for your skin. The natural saponification process results in about 75 per cent soap and 25 per cent glycerine, an emollient that softens, protects and heals the skin.

With a pH of around 8 or 9, cold-process soaps are gently alkaline and cleansing on the skin, which has a slightly acid pH, varying between 4.5 and 6.2. The alkali is sodium hydroxide (with a pH of around 14), which is mixed with water to create lye – a necessary ingredient in the creation of soap. Handmade soap is the neutral end-product of combining an alkali and several oils. Sodium hydroxide is a compound containing three elements – sodium, hydrogen and oxygen – and it is transformed or saponified during the soap-making process.

The acids are oils or fats (olive, coconut, almond and essential oils and shea butter). They are listed as sodium olivate, sodium cocoate, sodium almondate and shea butterate in the ingredients list on the label (known as the INCI – or International Nomenclature of Chemical Ingredients) because they are transformed into the sodium salts of the fatty acids during the soap-making process.

Take your time – you can't rush the process or be distracted. If you follow the next few steps, you will be able to make gentle, natural all-over hair and body soap bars safely and effectively.

I use silicone baking moulds as they make it so much easier to turn soaps out once set. You can use small shapes, or larger rectangular loaf shapes, which means you can slice the soap with a knife. These moulds can be easily found in cookware shops and online.

STEPS IN BOTANICAL SOAP-MAKING

1. Weighing ingredients and melting solid fats

First, weigh all the ingredients separately. In a saucepan, gently melt the solid fats (such as coconut oil or shea butter) with the liquid oils. This takes just a few minutes, then remove the pan from the heat and leave the mixture to cool to around 40°C (104°F). (You can use the bain-marie style water cooling method, lowering the bottom of the saucepan into a bowl of ice-cold water. Make sure that none of the cold water enters the saucepan.)

2. Making and cooling the lye

The lye solution is created by dissolving sodium hydroxide in cold water (creating an exothermic reaction that gives off heat) and setting it aside to cool until it also reaches 40°C (104°F). Make sure you wear rubber gloves and eye protection, and pour the sodium hydroxide into the water and not the other way around, to avoid splashing. Again, you can use a cold water bain-marie to speed up the cooling process.

3. Mixing and emulsifying

When the right temperature is achieved, the lye solution is gently mixed into the oils in the pan. Keep stirring with a spoon (or use a hand-held electric blender to speed up the process) until the mixture reaches what is known as the moment of trace. This is the transformative moment in the soap-making process and it requires focus and steady hands. With no fat molecules left at the sides, the now uniformly coloured mixture leaves a definite viscous trace on the surface – this tells you it is thoroughly emulsified. At this stage, it looks like a thin, pale custard, and you can stir in any additional ingredients, such as essential oils, flower petals, seeds, sea salt or seaweeds.

4. Moulding and setting

Transfer the soap mix into a jug and then pour it into a silicone mould or soap moulds. You can sprinkle extra ingredients, such as flower petals, seeds or seaweeds on the top of each soap.

The gel phase of soap-making is the heating stage of saponification. Once you pour your soap into its mould, the heat within the soap mix is concentrated in the centre, as the sides cool. The gel phase starts with the soap turning translucent in the middle and then spreading out to the edges.

The soaps should be left to set completely for at least 24–48 hours before being unmoulded onto trays and carried to curing racks (use any wooden or metal racks you have) lined with clean tea towels, where they can dry and cure at a cool ambient temperature. If you are making lots of different batches, label the different soap names, batch numbers and dates.

5. Curing

During the curing stage, which takes 30–40 days, the soaps become fully saponified and dry out, oxidizing in the air and lightening in colour. During this stage, air needs to circulate round the whole soap to ensure that no moisture remains. Turn the soap bars a couple of times during curing to ensure they dry evenly.

6. Storage

Once the soaps are fully cured and the moisture has completely evaporated, you can store your soaps in a plastic or cardboard box with a lid. Make sure you always label the container with the name of the soaps and the date made, especially if you are making several batches.

SAFETY NOTE

Wear goggles, rubber gloves and a long-sleeved shirt when handmaking soap to ensure the alkaline raw soap mixture does not splash in your eyes or on your skin. If it does, wash with copious amounts of water.

HYDROSOLS

Hydrosols, herbal waters and herbal distillates, are all water-based products of distillation, along with the essential oils.

Floral and herbal distillations are used as flavourings in food **and** are in topical skincare products such as toners and cleansers, or face and laundry mists. They can also be added to water, ice and lemon for a refreshing summer drink. Popular herbal distillates for skincare include rosewater, orange flower water and witch hazel. Rosemary, oregano and thyme are also favoured for their antimicrobial and antibacterial properties. Most herbs should be distilled in the summer before they flower, whereas roses can be picked in full bloom for their petals.

As hydrosols have antioxidant, antibacterial and acidic properties, they are a valuable bi-product of the distillation process. They are not sterile, however, so need to be kept in a sterilized bottle in the fridge.

Herbal distillations are often produced by the same process as essential oils. In the distillation process, essential oils float to the top of the water. Commercial skincare makers will use a copper alembic still to extract essential oils and hydrosols, by distillation, but to make your own hydrosols it's entirely possible to use a simple stovetop alternative – either a 20cm (8-inch) three-tier steamer pan set or a bamboo steamer inside a large saucepan.

INGREDIENTS

- 1 litre (1¾ pints) water
- Enough fresh herbs to fill one tier of a steamer (you can use a variety of plants, for example, rosemary, mint, lavender, rosehips, rose petals, thyme, sage)
- Ice cubes

METHOD

Pour the water in to the bottom pan, and place the plant material (herbs or flowers) in the second tier or steamer basket above it, ensuring that the water does not reach the base of the middle tier where your herbs sit. Ensure that the herb tier is quite full, leaving just 1cm (½in) free space at the top.

Select a heatproof bowl slightly smaller than the size of the steamer, one that will fit in the top tier, which is where the distillate will condense. On top of this tier, place an inverted lid topped with lots of ice cubes, or ice bags. When the rising steam distillate meets the ice barrier, it will condense back to liquid and flow down into the bowl beneath.

Heat the water to boiling point, then simmer for around 20 minutes. The steam will rise, extracting the oils from the plants in the second tier, before condensing into a hydrosol in the bowl. A very small amount of essential oil will float to the top of the hydrosol, but it will probably be too small to collect separately, so I just use it with the hydrosol, combined. Decant your hydrosol into sterilized glass bottles, seal, label and date. Store in the fridge, where it will keep for 3 months.

ENFLEURAGES

An enfleurage is a traditional technique that is widely used in the natural perfumery market for extracting essential oils and perfumes from plant material using natural fats, such as plant oils, butters or animal tallow.

I use the technique to extract the natural antioxidant and nutrient properties of the plants I find, including foraged leaves, berries, wood, herbs and flowers. It can also be used for extracting the perfume of jasmine, roses, lilacs and lilies.

For this formulation, you can forage and cut widely, as the wider range of plants available makes for a richer and more beneficial skin oil.

The method here for enfleurage straddles both cold and hot techniques – using coconut oil, which is solid at room temperature, but melts when warmed.

In the world of enfleurage-making, the term 'menstruum' refers to the solvent or base that is being used to extract the plant properties, such as oil, alcohol, glycerine, honey, etc. 'Marc' is the spent plant material at the end, which is simply disposed of on the compost heap.

Just like the macerated oil, you can use your enfleurage oil as an all-over face, hand and body moisturizer and massage oil, both day and night, especially after a bath or shower.

INGREDIENTS TO MAKE 250ML (9FL OZ) ENFLEURAGE SKIN OIL

- 200g (7oz) coconut oil (organic, refined if possible)
- 200g (7oz) freshly picked seasonal herbs, leaves, flowers and berries

METHOD

Gently melt the coconut oil and pour into a small, shallow, heatproof metal or ceramic baking tray.

Distribute half the herbs, leaves, flowers and berries over the oil, using the back of a spoon to submerge the plant material.

Cover the tray with a clean cloth and leave in a cool spot for 24 hours for the coconut oil to cool and solidify.

The next day, place the tray on a sunny window sill or in a warm, switched-off oven, just long enough for the oil to melt. Using a slotted spoon, remove the plant material from the oil and dispose of it.

Repeat the process again with the remaining herbs, leaves, flowers and berries and leave to cool for a further 24 hours.

The following day, repeat the melting process and then remove the plant material. Strain the coconut oil into a clean jug and then pour into a sterilized bottle, seal, label and date. It will keep for 3 months in a cool, dark cupboard.

TINCTURES

Tinctures are concentrated, alcohol-based extracts of plant material. You can use vodka, which is tasteless, or gin or brandy and dried berries, leaves, hips, flowers, bark, roots or rhizomes of plants – the ethanol in the alcohol is effective at extracting the properties from the plants. To begin with, you can start with just 2–3 drops of a herbal tincture, maybe taken with a warm herbal tea to increase the potency of its skin benefits. This can be built up to 1 teaspoon with meals twice or three times a day. Tinctures can also be used in salves and balms.

If you prefer to use vegetable glycerine, instead of alcohol, you can create a glycerite, which is a vegetable-based plant extract. Glycerine is an efficient skin humectant, drawing water from the air to the skin, and it is also a good preservative. Use a ratio of 55 per cent glycerine to 45 per cent plants and ensure that all the plant material is fully covered in a sterilized jar. Leave the glycerite to develop in a cool, dark place for 4–6 weeks, shaking the jar regularly and ensuring the plant material is always submerged.

INGREDIENTS TO MAKE 800ML (28FL OZ) OF TINCTURE

50g (1¾oz) dried herbs or dried flowers
800ml (28fl oz) alcohol – vodka, gin or brandy

METHOD

Sterilize a 1-litre (1¾-pint) jar and lid in hot water. Dry with a clean tea towel.

Place the dried herbs or flowers in the jar and cover with the alcohol. Seal the jar tightly and shake to ensure that all of the plant material is submerged.

Place in a cool, dark place for 3 weeks and give the jar a few good shakes from time to time.

Strain the ingredients through a clean muslin cloth into a jug. Discard the plant material and pour the tincture into a sterilized amber or blue glass bottle. Seal, label and date the bottle and use within 12 months.

APPLE AND CRAB APPLE CIDER VINEGAR

INGREDIENTS FOR A 1-LITRE (1¾-PINT) JAR

- 4 large eating or cooking apples or 15–20 small crab apples, chopped into medium-sized pieces (peel and cores included)
- Approximately 500ml (18fl oz) freshly drawn tap water or enough to cover all the apples
- 25g (1oz) raw cane sugar

Apple cider vinegar (ACV) is rich in malic acid, which helps boost the immune system – I drink a spoonful a day – and it also makes an effective, astringent toner, exfoliating the skin and removing dead skin cells. It is a shine-inducing toner for hair, too, keeping dandruff at bay. You can use eating apples, cooking apples or foraged crab apples here and include the peel and cores. I always wash and dry the apples before use. Make this with raw cane sugar for its intense colour, flavour and fermentation profile, with a deeper molasses and caramel taste.

Sterilize a 1-litre (1¾-pint) jar and lid in hot water. Dry with a clean tea towel.

Place the chopped apples in the jar.

Warm the water in a small pan, add the sugar and stir to dissolve completely. Pour enough of the water-sugar solution into the jar to cover the apples completely, right to the top, allowing a 1cm (½in) gap at the rim.

Cover the jar with a clean linen square and tie it with string. During fermentation gases will be released, therefore you can place the lid on top but don't screw it tightly.

Place the jar in a cool, dark cupboard, larder or basement for 4 weeks.

Strain the ACV into a clean jug and discard the apple pieces. Pour the vinegar into a sterilized glass bottle and seal with a cork or screw top. Label and date the bottle. This keeps for 6 months in a cool cupboard.

WINTER FORMULATIONS

FOUR SALTS, SAGE & SEAWEED BATH SALTS

INGREDIENTS FOR A 500G (1LB 2OZ) JAR

- 100g (3½oz) Dorset sea salt (or any British sea salt, such as Cornish, Halen Mon, Blackthorn or Isle of Skye)
- 100g (3½oz) rock salt
- 100g (3½oz) Epsom salts
- 100g (3½oz) Himalayan salt
- 50g (1¾oz) dried sage leaves, very finely cut
- 50g (1¾oz) dried seaweed (Bladder wrack), very finely cut
- 10–12 drops of sage essential oil

This is one of my favourite bath salt recipes for a warming and therapeutic pick-me-up. The minerals in the different salts and seaweed relax and cleanse while helping to detox and de-stress the body. Sage leaves and essential oil are antiseptic, cleansing and deodorizing. They also help stimulate the skin and have anti-ageing properties. To use, simply run a warm bath and spoon 4 heaped tablespoons of the bath salts into a clean muslin bag with a string closure. Close the top and immerse the bag in the water. Soak for 15–20 minutes.

Sterilize a 500g (1lb 2oz) glass jar and lid with hot water and soap. Rinse, dry and set aside.

In a clean bowl, mix together the salts, sage, seaweed and essential oil.

Spoon the bath salts into the jar. Seal and label and use as above.

SEA SALT BATH SCRUB WITH ROSE-SCENTED GERANIUM

INGREDIENTS FOR 1 TREATMENT

50g (1¾oz) sea salt
50ml (2fl oz) almond oil
2 heaped tbsp finely cut fresh rose-scented geranium leaves
5–10 drops of geranium essential oil

Gentle exfoliation is an important part of the face and body skincare regime, especially in winter. This formulation will slough off dead skin cells, improve circulation, cleanse and tone the skin while moisturizing the face and body and helping you to detox, refresh and relax. Generally, when using scrubs, avoid the delicate eye area.

In the bath or shower, take a small amount of scrub and work using circular motions into your skin, from face to toe, giving particular attention to your feet, knees, elbows, hands, neck and décolleté. Scrub until your skin feels tingly and becomes pink from increased circulation. Rinse thoroughly with warm water. Pat dry with a clean towel.

In a clean bowl or jar, mix all the ingredients thoroughly.

Leave to stand, covered, for a few minutes so that the fragrance is imbued throughout the mix. Use within 3 days.

DRIED CALENDULA, LAVENDER & ROSEMARY OIL MACERATION

INGREDIENTS FOR A 250ML (9FL OZ) BOTTLE

200g (7oz) mixed dried calendula flowers and lavender and rosemary stems and flowers

200ml (7fl oz) mixed botanical oils (a mix of rapeseed, olive, sunflower, almond, hemp, borage or flaxseed)

I use more all-over skin oil in winter than at any time of year. After a warm sea salt bath and scrub, I apply this face and body oil with circular movements. It's non-greasy as the skin absorbs it all immediately, and a little goes a long way.

Using the maceration process described on page 109 enables you to use the late summer and early autumn herb and flower harvest to create a fragrant and moisturizing oil that will help keep your skin soft and supple during the colder months. A bottle of this oil also makes a wonderful gift, beautifully wrapped and labelled.

Sterilize a 500ml (18fl oz) jar and lid in hot water and dry.

Place the flowers and herbs in the jar and top with the oil. Close the lid tightly and shake the jar a few times. Leave on a sunny window sill for 4 weeks or in a cupboard for 6 weeks. Shake the jar regularly.

When it is ready to use, strain the oil through a sieve into a clean jug and discard the herbs and flowers. Pour into a sterilized bottle, seal, label and date. Oil macerations keep, sealed, for up to 12 months in a cool, dark cupboard or bathroom cabinet.

WINTER SKINCARE TEA

INGREDIENTS TO MAKE 2 CUPS

1 litre (1¾ pints) kettle-hot water
1 tsp white, green or rooibos tea leaves
2 tsp dried rosehips
2 tsp dried unsweetened berries (cranberries, mulberries, raspberries)
Slices of dried or fresh apple (with peel)
Slices of clementine or lemon (with peel)

Herbal teas are an excellent way of ensuring that the skin has plenty of hydration. In winter we tend to live indoors, enveloped in warm clothes, in centrally heated homes, where our diets may consist mainly of comforting foods, such as carbohydrates, animal fats and baked goods. Staying hydrated is one of the fundamentals of natural skin biome supportive skincare and herbal teas are a flavoursome and fragrant way of ensuring we drink enough water every day. In addition to white, green or rooibos teas, which are high in antioxidants, I add dried berries, herbs and flowers to teas throughout the day. In my workshop and kitchen, there are jars of dried herbs, flowers, citrus peels and spices, such as cinnamon, ginger and cloves, ready for making seasonal herb teas which I either drink warm, with a slice of lemon or lime, or cold with ice and mint leaves.

Use 500ml (18fl oz) of the hot water to warm the teapot and cups. Pour the water away.

Add the tea leaves, rosehips, berries, apple and citrus to the warm pot, and pour in the remaining 500ml (18fl oz) of hot water.

Infuse the herbal tea for 3–4 minutes, then strain into the cups.

EARLY SPRING FORMULATIONS

—

CALENDULA, PEPPERMINT & POPPY SEED SOAP

INGREDIENTS TO MAKE APPROXIMATELY 1KG (2LB 4OZ) OF SOAP

300ml (10fl oz) calendula macerated olive oil (see page 109), strained
200g (7oz) coconut oil
200ml (7fl oz) rapeseed oil
170g (6oz) sodium hydroxide
225ml (8fl oz) water
20g (¾oz) dried calendula petals
20g (¾oz) poppy seeds
2 tbsp peppermint essential oil

You should follow the botanical soap-making method outlined on pages 111 and 112 for this formulation before you begin.

Put the macerated olive oil with the coconut oil and rapeseed oil in a saucepan and gently melt. Leave the mixture to cool to around 40°C (104°F).

Pour the sodium hydroxide into the water in a bowl and mix well. Leave to cool to 40°C (104°F).

When the right temperature is achieved, add the lye solution to the oils and mix thoroughly.

Keep stirring until the mixture reaches the moment of trace, where the emulsified mixture leaves a definite viscous trace on the surface. Stir in half the dried petals and poppy seeds and the essential oil.

Pour the soap mix into a jug and then pour into a silicone mould or several cake moulds. Sprinkle the remaining half of dried petals and poppy seeds on the top of each one.

Leave the soaps to set completely for at least 48 hours, then unmould onto trays and leave to dry and cure for 30–40 days on curing racks lined with clean tea towels.

CRAB APPLE (OR APPLE) CIDER VINEGAR, LEMON & ORANGE FLOWER TONER

INGREDIENTS TO MAKE 200ML (7FL OZ)

- 50ml (2fl oz) distilled water
- 50ml (2fl oz) crab apple (or apple) cider vinegar
- Juice of 1 lemon
- 50ml (2fl oz) orange flower water

Malic and citric acid are effective exfoliators, and you get both with these ingredients. This toner is an effective way to support your skin microbiome in the process, stimulating the circulation as well as regulating the skin's naturally acidic pH.

Combine all the ingredients in a clean jug and mix well. Decant into a sterilized glass bottle. Seal, label and date. Refrigerate and use within 3 months.

Apply with a linen or muslin cloth or cotton wool pads and gently wipe over the skin.

LEMON BALM ALL-OVER BODY BALM

INGREDIENTS TO MAKE 40G (1½OZ)

- 20g (¾oz) cacao butter
- 20g (¾oz) coconut oil
- 10 drops of lemon balm tincture (see page 119)
- 10 drops of lemon essential oil

Balms are a simple way to help nourish and protect the skin – particularly those areas that are prone to extreme dryness, such as lips, elbows, heels and knees. Cacao butter is made from the pressed oil of the cacao bean. Lemon balm brings its healing and antibacterial powers to dry skin, cold sores, poorly healing wounds, infections, and minor insect bites and stings.

Once cooled and set solid in a sterilized jar, you can take your balm with you wherever you go travelling, in your bag or keep it in the car.

Put the cacao butter and coconut oil in a bowl over a saucepan of simmering hot water, bain-marie style. Stir until melted.

Mix in the lemon balm tincture and lemon essential oil.

Pour into a small, sterilized jar or tin to set. Seal, label and date. It will keep, unopened, for 3–6 months. Once opened, use within a month.

CLEAVERS, HONEY & CLAY FACE MASK

INGREDIENTS FOR 1 FACIAL APPLICATION

3 tbsp cleavers water infusion (see page 108)
1 tsp clear honey
2 tbsp clay powder
2 drops of rose essential oil

Natural clay minerals help to draw impurities from the skin and cleanse it. A foraged cleavers infusion makes the perfect astringent tonic for a spring cleaning of the skin and, combined with honey and rose essential oil, you will feel revived and refreshed with this soothing mask.

Mix together the cleavers infusion and honey in a bowl.

Add the clay powder by sprinkling it on the liquid and stirring well. Strain the mix through a sieve to distribute the powder evenly. Add the rose essential oil and mix.

Apply the clay mask to freshly cleansed skin, avoiding the area round the mouth and eyes.

Leave the clay mask on your face for 10 minutes. Relax.

Rinse off the mask with warm water and a face cloth and pat dry with a towel.

WHITE NETTLE, COMFREY, DRIED SAGE AND MINT HAIR & SCALP TONIC

INGREDIENTS TO MAKE A 300ML (10FL OZ) TONIC RINSE – ENOUGH FOR 1 RINSE

300ml (10fl oz) distilled water
2 small sprigs of fresh white nettle
2 small sprigs of fresh comfrey
2 tbsp dried sage
2 tbsp dried mint
2 tbsp apple cider vinegar
6 drops of peppermint essential oil

Scalp health and hygiene is very important for the care and condition of your hair. You just need a homemade botanical hair and body soap to wash your hair and then rinse it thoroughly with warm water – thereby avoiding the build-up of chemicals on the scalp and hair cuticles which make your hair look dull and lacking shine or lustre. The apple cider vinegar in this natural, plant-based tonic rinse helps to cleanse, detoxify and balance scalp pH and remove any chemical build-up, which may be the cause of a dry, flaky scalp and dandruff. Use this tonic rinse at least once a month.

Tap water may contain a number of impurities and so using distilled water will ensure a cleaner rinse.

First make an infusion – boil the distilled water in a saucepan or clean kettle that has no limescale in it.

Place the nettle, comfrey, sage and mint in a heatproof jug and pour the hot water over them.

Leave the infusion to steep for 10 minutes. Strain through a sieve into a clean jug.

Mix in the apple cider vinegar and peppermint essential oil.

Apply the warm tonic rinse after showering to your wet, clean scalp and hair and massage gently. Comb through your hair, from roots to ends. Rinse and towel-dry your scalp and hair.

SPRING FACIAL TEA & STEAM – DANDELIONS, CLEAVERS, WHITE NETTLES

Three herbs with three applications

INGREDIENTS TO MAKE 1 FACIAL STEAM

Leaves and flowers of 3 fresh dandelions
Leaves and flowers of 3 fresh cleaver stems
Leaves and flowers of 3 fresh white nettles
4 drops of grapefruit essential oil

Spring is the time of renewal and cleansing and foraged herbs make for vital skincare detox, both internally and externally. During the early spring foraging season, I make sure I collect plentiful amounts of dandelion, cleavers and white nettle flowers and leaves to use fresh for herbal infusions, to drink as a tea without the essential oil, and to create a facial steam, including an essential oil.

Tea made from dandelions, cleavers and white nettles helps support the skin biome through the diuretic effectiveness of these herbs, reducing the puffiness of winter skin and improving circulation. While you are waiting for your herbal tea to cool to drinking temperature, you can enjoy a spring facial steam to get your skin glowing. Add a few drops of your favourite essential oil – the invigorating fragrance of grapefruit is particularly uplifting at this time of year.

Put the dandelions, cleavers and nettles in a bowl and cover with hot water from a kettle.

Allow to cool slightly and place a towel over your head to create a steam tent over the bowl.

Breathe deeply for 10 minutes and relax as your face is steamed.

Once the steam is finished, strain the resulting skincare liquid and add to your next bath for a relaxing, skin-nourishing soak, or use as a facial toner or rinse your hair with it.

LATE SPRING FORMULATIONS

—

SEAWEED BATH SOAK

INGREDIENTS FOR
1 BATH SOAK

30g (1oz) dried seaweed (I used bladder wrack), coarsely chopped
200g (7oz) jumbo oats
100g (3½oz) bicarbonate of soda

Seaweeds have been used for millennia for their skin rejuvenating qualities. These sea algae are packed with natural antioxidants and essential vitamins that can turn a simple bath into a liquid skin-enhancer. Seaweed baths can help repair the skin, reduce the signs of ageing and cellulite, and ease arthritis and eczema.

Bicarbonate of soda helps to soften the water and is highly cleansing for the skin. Oats added to a bath bring moisturizing, soothing, inflammation relief to the skin. They provide a useful make-at-home remedy you can use to treat a variety of skin conditions, from psoriasis to eczema.

Place the seaweed and oats in a muslin bag and soak in a bath of hot water. The seaweed will naturally rehydrate as the hot water causes the release of the plant's alginate, a vitamin- and mineral-rich gel. Give the bag a good squeeze under the water to release all the nutrients.

Sprinkle the bicarbonate of soda in the bath water.

Add cooler water to ensure a comfortable bath temperature before entering and then relax for a good 10–15 minutes, rubbing the bag against your skin.

Take care when you leave the bath as the gel is slippery.

SPRING SKIN OIL ENFLEURAGE

INGREDIENTS TO MAKE 200ML (7FL OZ) ENFLEURAGE SKIN OIL

- 200ml (7fl oz) mixed botanical oils – borage, grapeseed, apricot, hemp, flaxseed, almond, rosehip, rapeseed, sunflower are all appropriate
- 100–200g (3½–7oz) freshly picked spring herbs, leaves and flowers – comfrey, dandelion, white nettles, borage, calendula, primroses, mint, lemon balm, jasmine, honeysuckle, roses, etc.

You should follow the enfleurage method (see page 116) for this formulation before starting.

Pour the oils into a small, shallow enamel or ceramic dish and mix.

Place half the herbs, leaves and flowers in the oil, using the back of a spoon to submerge the plant material.

Cover the dish with a clean cloth and leave for 24 hours in a cool corner.

The next day, using a slotted spoon, remove the plant material from the oil and dispose of it on the compost heap.

Repeat the process with the remaining herbs, leaves and flowers and the same oil, using a spoon to submerge the plant material and again leave for a further 24 hours in a cool place.

The following day, strain the oil into a clean jug and then pour into a sterilized bottle, seal, label and date. It will keep for 3 months in a cool, dark cupboard.

PEPPERMINT & ROSEMARY FOOT CREAM

INGREDIENTS
TO MAKE 100g (3½oz) OF FOOT CREAM

- 3 tbsp almond oil
- 2 tsp cacao butter
- 2 tsp beeswax pellets
- 2 tsp emulsifying wax
- 2 tbsp peppermint and rosemary infusion (see page 108)
- 6 drops of peppermint essential oil
- 6 drops of rosemary essential oil

The feet are often the most neglected parts of the body. The skin, particularly on our heels, is often prone to hardening and cracking from dryness, which can lead to discomfort or even pain.

Once a week, I treat myself to a foot bath with warm water, sea salt, bicarbonate of soda and Epsom salts. I then exfoliate my feet all over with a pumice stone, brush and loofah and I carry out a mini pedicure to keep my feet in good condition. Little and often is the best route to good foot care and hygiene.

Peppermint helps relieve aches and pains (as does common garden mint – or any mint variety) and the menthol fragrance is lively and refreshing. Rosemary is antibacterial and helps improve the circulation.

Create an emulsion of the almond oil, cacao butter and beeswax by placing them all in a heatproof bowl over a saucepan of gently simmering water. Stir until the beeswax has melted and then remove from the heat.

Heat the herb infusion in a saucepan to 80°C (176°F). Off the heat, dissolve the emulsifying wax in it.

Add the infusion to the almond oil mixture and, using a whisk or hand-held blender, mix until the emulsified cream is completely smooth. Continue to mix while the mixture cools.

Add the essential oils, mix again and then decant into a small, sterilized jar with a lid; seal, label and date. Store in the fridge for up to 4 weeks.

To apply, massage the cream into clean feet in a circular motion, concentrating on the heels.

SAGE, ROSEMARY & THYME ALL-OVER BODY WASH

INGREDIENTS FOR 250ML (9FL OZ) BOTTLE OF HAIR & BODY/HAND WASH

150ml (5fl oz) concentrated, unscented Castile liquid soap
2 long sprigs each of dried sage, rosemary and thyme
6 tbsp hot water
1 tsp each of sage, rosemary and thyme essential oils

There is nothing more invigorating and refreshing at the beginning or end of the day than an all-over herbal hair and body wash. Sage, rosemary and thyme are a powerful combination, bringing anti-inflammatory, antioxidant, antibacterial and cleansing benefits to a warm shower or bath.

This is my multi-functional wash – I keep pump dispenser bottles filled with it in the bathroom as well as by the kitchen sink and use it for hand washing as well. It has a green and pleasant smell, the fragrance of summer, slightly medicinal but completely delicate.

In this recipe, I am using organic, concentrated Castile liquid soap – this is quite a challenging soap to make from scratch and there are lots of really good commercial alternatives: one I recommend is Dr. Bronner's concentrated, unscented Castile liquid soap. Although this soap is named after the former Spanish kingdom of Castile, it was originally made as a hard soap in the Levant, using laurel and olive oils. Laurel oil was also abundant in the Castile region, although the main components of Castile liquid soap today are normally olive oil, sunflower oil, distilled water and lye.

Pour the Castile liquid soap into a clean jug.

In a separate jug, make an infusion of the dried herbs in the hot water. Strain once infused.

Pour the herb infusion and the essential oils into the liquid soap jug and mix thoroughly using a whisk.

Pour the soap mix into a sterilized bottle with a pump dispenser or lid. This will keep for 6 months.

ROSE & ST JOHN'S WORT TONER

INGREDIENTS TO MAKE 200ML (7FL OZ) TONER

180ml (6¼fl oz) distilled water
5 heaped tsp fresh rose petals
3 heaped tsp fresh St John's Wort flowers and leaves

Rose petals and St John's Wort flowers and leaves make a refreshing and astringent toner, effective at calming and balancing the skin after cleansing and removing all final traces of make up or impurities.

Make an infusion (see page 108) with the water, petals, flowers and leaves. Leave to cool.

Strain into a sterilized glass bottle. Seal, label and date.

Refrigerate and use within 3 months.

Apply with a linen or muslin cloth or cotton wool pads and gently wipe over the skin.

EARLY SUMMER FORMULATIONS

—

EVENING PRIMROSE & ROSE MOISTURIZER

INGREDIENTS TO MAKE 100G (3½OZ) OF MOISTURIZER

- 2 tsp cacao butter
- 4 tsp evening primrose oil
- 4 tsp rosehip oil
- 2 tsp beeswax pellets
- 2 tsp emulsifying wax
- 2 tbsp rose and evening primrose flower infusion (see page 108)
- 8 drops of rose essential oil

This moisturizer is for everyday use on the face, hands and décolleté. It is creamy, rich and highly effective for ageing skins, especially during the dry and warmer months. I make a jar of this every couple of weeks, as I use it every day and the fragrance is very uplifting.

Place the cacao butter, evening primrose oil and rosehip oil in a heatproof bowl over a saucepan of simmering water and gently heat until the wax melts. Remove from the heat.

Dissolve the emulsifying wax in the hot flower infusion.

Add the infusion to the oil mixture and, using a whisk or hand-held electric blender, whisk continuously until the cream is smooth. Continue to mix as the mixture cools.

Add the essential oil, mix again and then decant into a small, sterilized jar or tin with a lid. Seal, label and date. Refrigerate and use within 3 months.

To apply, massage the cream onto your cleansed face in circular motions. Pat using your fingers very gently around the eye and mouth areas. Apply the moisturizer to your neck and décolleté with upward and outward motions.

PEPPERMINT & GERANIUM FACE AND BODY SPRITZ

INGREDIENTS TO MAKE A 200ML (7FL OZ) SPRAY

- 1 heaped tbsp fresh peppermint or garden mint leaves
- 1 heaped tbsp fresh scented pelargonium leaves
- 50ml (2fl oz) hot water
- 40ml (1½fl oz) rosewater
- 5 drops of geranium essential oil

A botanical spritz will help keep you refreshed and cool during the summer heat, both day and night. The cooling element of menthol in the peppermint and the uplifting fragrance of pelargonium and geranium are key ingredients for staying fresh and smelling clean. I also use my botanical spritz on pillows, pyjamas and wrists before bedtime – if you suffer from sleepless nights in summer, this will alleviate discomfort and heat.

Make an infusion of the peppermint or garden mint and pelagonium in the hot water. Leave to cool.

Strain the infusion into a sterilized spray bottle. Add the rosewater and geranium essential oil. Seal and shake.

Keep in the fridge and use within 3–5 days.

BORAGE & CLOVER NIGHT CREAM

INGREDIENTS TO MAKE 100G (3½OZ) OF CREAM

- 2 tsp cacao butter
- 3 tbsp borage seed oil
- 2 tsp beeswax pellets
- 2 tsp emulsifying wax
- 2 tbsp borage and clover flowers infusion (see page 108)
- 1 tsp dried borage flowers
- 1 tsp dried clover flowers

The hours when the body is asleep are an important regenerative time for the skin, which is why sleep is so important for good skincare. Overnight, skin undergoes repair and restoration, which is why plants that are high in antioxidants and vitamins, such as borage and crimson clover, are beneficial for their anti-ageing and moisturizing properties.

Put the cacao butter, borage seed oil and beeswax in a heatproof bowl over a saucepan of simmering water and gently heat until the wax melts. Remove from the heat.

Dissolve the emulsifying wax in the hot flower infusion.

Add the infusion to the oil mixture and, using a whisk or hand-held electric blender, whisk continuously until the cream is smooth. Continue to mix as the mixture cools.

Using a pestle and mortar, grind the dried borage and clover flowers to a powder and mix into the cream. Decant into a small, sterilized jar or tin with a lid. Seal, label and date. Refrigerate and use within 3 months.

Apply before bedtime, massaging the cream onto your cleansed face in circular motions. Pat using your fingers very gently around the eye and mouth areas. Apply the moisturizer to your neck and décolleté with upward and outward motions.

SUMMER SKIN OIL MACERATION – CALENDULA, BORAGE & ST JOHN'S WORT FACE & BODY OIL

INGREDIENTS FOR A 250ML (9FL OZ) BOTTLE

200g (7oz) mixed dried calendula, borage and St John's Wort flower heads
200ml (7fl oz) borage seed oil

Calendula, borage and St John's Wort are filled with antioxidant, anti-inflammatory and anti-ageing properties. During the summer months, my garden overflows with their flowers and I make huge jars of their oil macerations, which last the rest of the year. The oil can be applied all over the face, body and hair, to help keep skin moisturized as well as providing relief for any bruises, aches and pains. Gentle, effective and not greasy, it can be used during the day and at night.

Sterilize a 500ml (18fl oz) jar and lid in hot water and dry.

Place the flower heads in the jar and top with the oil. Close the lid tightly and shake the jar a few times. Leave on a sunny window sill for 4 weeks or in a cupboard for 6 weeks.

Shake the jar regularly. When you want to use it, strain the oil through a sieve into a clean jug and discard the herbs. Pour into a sterilized bottle, seal, label and date. Oil macerations keep for up to 12 months in a sealed jar in a cool, dark cupboard or bathroom cabinet.

CHAMOMILE & COMFREY BATH INFUSION

INGREDIENTS FOR 1 BATH

10 dried chamomile flower heads
10 dried lavender flower heads
10 dried comfrey flower heads and 10 leaves
1 litre (1¾ pints) kettle-hot water
3 tbsp sea salt
3 tbsp apple cider vinegar (see page 120)
5 drops of lavender essential oil
5 drops of chamomile essential oil

A herbal bath infusion is an excellent way of creating a relaxing and cleansing home spa. Dried chamomile and lavender flowers have soporific properties and have been used for centuries to aid sleep and relaxation. Comfrey contains allantoin, a compound that helps to soothe skin, contributes to skin renewal, reduces inflammation and acts as an emollient.

Using dried herbs in your infusion will nurture, fragrance and refresh you, without leaving a trace in the bath tub because the mixture is strained before use.

Make a concentrated infusion by placing all the plant material in a jug or teapot and pour the kettle-hot water on top. Stir and set aside, covered, for 5 minutes.

Strain the infusion into a clean jug. Add the salt, vinegar and essential oils and stir again. Add to your warm bath water and use a face flannel to wash your face and body with the infusion dilution. Immerse yourself for 15–20 minutes.

SUMMER HERBAL TUSSIE-MUSSIE BATH FIZZ

INGREDIENTS TO MAKE 1 TUSSIE-MUSSIE

One stem of each of the flowers and herbs growing in the skincare garden

A tussie-mussie is a small flower bouquet. The unusually named tussie-mussie dates from Victorian times when these bouquets were carried as a popular fashion accessory and it was common to include herbs, both for their scent and for their healing properties. During the summer months, there is a bounty of fresh herbs and flowers in my skincare garden, and I use tussie-mussies in several ways.

Remove any lower leaves and cut all the stems to the same length.

Tie the posy with string.

Use fresh by hanging it under the hot water tap in the bath or the shower head when you shower. Or place tips of herbs and flower petals in a bowl of hot water to make a facial steam and, with your head covered by a towel, hold your face just above the bowl.

Hang a small tussie-mussie upside down in a warm place, like an airing cupboard or a warm kitchen. Once dried, the herbs and flowers can be used for herbal teas, or to add to bath salts or salt scrubs, along with a few drops of essential oil.

INGREDIENTS TO MAKE 4 TUSSIE-MUSSIE BATH FIZZES

100g (3½oz) bicarbonate of soda
30g (1oz) citric acid
30g (1oz) Epsom salts
Petals and leaves from 1 small dried tussie-mussie
10 drops of lavender essential oil
A little tap water

Mix together the bicarbonate of soda, citric acid and Epsom salts in a large bowl.

Use your fingers to break up the dried flowers and herbs from the tussie-mussie, sprinkling them into the bowl.

Add the essential oil. The mix may start fizzing, so stir quickly to blend it all.

Using a water spray, spray just enough water into the mix so that it starts to clump together – don't over wet it. You will need to use your hands to mix and form four equally sized ball shapes. They will flatten as they dry.

Leave the bath fizzes to set overnight. Store in an airtight jar and use within 2 months.

To use, run a warm bath, then place a fizz ball in it until completely dissolved.

LATE SUMMER FORMULATIONS

YARROW, ST JOHN'S WORT & EVENING PRIMROSE FACE TONIC

INGREDIENTS

1 litre (1¾ pints) water

Fresh mixed leaves and flowers of yarrow and St John's Wort, and evening primrose – enough to fill the second tier of a steamer

Ice cubes

Do read the hydrosol-making method outlined in the introduction to this chapter (see page 115) for this formulation before starting.

This face toner is created from the steam distillation of the aerial parts (leaves and flowers) of yarrow, St John's Wort and evening primrose.

Used together, the astringent and anti-inflammatory properties of these plants make for a soothing, calming and refreshing toner for summer days. Keep a small bottle in the fridge and use in the morning after cleansing, after an evening bath to help prepare your skin for sleep, and during travel to keep you calm and cool. It makes a refreshing floral mist sprayed on pillows, handkerchiefs, or on your wrists and neck.

Pour the water in to the bottom pot of the steamer, and place the plant material in the second tier above it. Make sure that the plant material tier is quite full, leaving just 1cm (½in) free space at the top.

Select a heatproof bowl slightly smaller than the size of the steamer, one that will fit in the top tier, which is where the distillate will condense. On top of this tier, place an inverted lid topped with lots

of ice cubes, or ice bags. When the rising steam distillate meets the ice barrier, it will condense back to liquid and flow down into the bowl beneath.

Heat the water to boiling point, then bring the temperature down to a simmer for around 20 minutes. The steam will rise, extracting the oils from the plants in the second tier, before condensing into a hydrosol in the bowl. A very small amount of essential oil will float to the top of the hydrosol, but it will probably be too small to collect separately, so I just use it with the hydrosol, combined.

Decant the hydrosol into sterilized glass bottles, seal with a lid, label, date and store in the fridge, where it will keep for 3 months.

LEMON BALM, LIME & THYME FLOWER HAND SANITIZER SPRAY

INGREDIENTS FOR 250ML (9FL OZ) SPRAY BOTTLE

4 small sprigs of fresh lemon balm, stems and leaves
4 small sprigs of fresh thyme, flowers and leaves
Zest and juice of 2 limes
200ml (7fl oz) clear alcohol (vodka)
10 drops of Castile liquid soap
10 drops of lemon essential oil
10 drops of thyme essential oil

This plant-based hand sanitizer is kind and gentle on the skin, but the antibacterial, antimicrobial and antiviral properties of lemon balm, thyme, citrus and essential oils are effective at ensuring that your hands are thoroughly clean when you cannot access soap and water.

I carry a small spray bottle in my bag, along with a linen cloth or tea towel – it can also be sprayed on shopping trolley handles, door handles or table counters when you are out and about or travelling.

Place the lemon balm, thyme and lime zest and juice in a 250ml (9fl oz) jar and cover with the clear alcohol to within 1cm (½in) of the rim of the jar and ensure that all the plant material is completely covered. Seal with the lid and label the jar.

Leave the jar in a cool, dark place for 3–4 weeks, then strain into a clean jug.

Add the liquid soap and the lemon and thyme essential oils. Mix well and decant the mixture into a sterilized spray bottle with a pump. Seal, label and use within 3 months.

ROSEMARY, POPPY SEED & SUGAR GARDENER'S SCRUB

INGREDIENTS TO MAKE 1 SCRUB

- 1 tbsp cane sugar
- 1 tbsp poppy seeds
- 1 tbsp finely chopped fresh rosemary needles
- 1 tbsp almond oil
- 1 tbsp sunflower oil
- 10 drops of rosemary essential oil

I regularly use a cane sugar scrub to ensure that my hands and nails are thoroughly cleaned from gardening grime and grit.

The poppy seeds and sugar granules in this scrub are gentle but effective at exfoliating dead skin cells, making your hands soft and nourished. As a natural source of glycolic acid, cane sugar encourages skin cell renewal and it is also a humectant. Unlike microplastics, both seeds and sugar can be rinsed down the drain without harming waterways or wildlife.

Rosemary is an invigorating and warming herb, naturally antiseptic and effective in soothing tired hands.

Mix together all the ingredients in a small bowl or jar, and it's ready to use.

Simply wash your hands with a botanical soap (see page 111) and rinse with warm water.

Apply the scrub to your wet hands, rubbing it gently and repeatedly all over your hands, wrists, nails and cuticles for 2–3 minutes.

Rinse thoroughly with warm water and dry with a clean towel.

Variation: Use poppy seeds mixed with coconut oil as a gentle but effective hand scrub to remove impurities and built-up grime.

LAVENDER & SEA SALT SOAP

INGREDIENTS TO MAKE APPROXIMATELY 1KG (2LB 4OZ) OF SOAP

200g (7oz) coconut oil
100g (3½oz) shea butter
200ml (7fl oz) lavender macerated olive oil (see page 109), strained
200ml (7fl oz) hemp oil
110g (3¾oz) sodium hydroxide
230ml (8fl oz) water
50g (1¾oz) sea salt
20g (¾oz) dried lavender flower heads
30ml (1fl oz) lavender essential oil

You should follow the botanical soap-making method outlined in the introduction of this chapter (see pages 111 and 112) for this formulation before starting.

Weigh all the ingredients separately. In a saucepan, gently melt the coconut oil and shea butter with the mascerated olive and hemp oils and then leave the mixture to cool to around 40°C (104°F).

Pour the sodium hydroxide into the water in a bowl and mix well. Leave to cool to 40°C (104°F).

When the right temperature is achieved, add the lye solution to the oils and mix thoroughly.

Keep stirring until the mix reaches the moment of trace, where the emulsified mixture leaves a definite viscous trace on the surface. Mix in the sea salt, half the dried flower heads and the essential oil by hand.

Pour the soap mix into a jug and then pour into a silicone mould or several cake moulds. Sprinkle the remaining half of the dried flowers on the top of each soap.

Leave the soaps to set completely for at least 48 hours, then unmould onto trays and leave to dry and cure for 30–40 days on curing racks lined with clean tea towels.

SUMMER HERBAL ICED TEA

INGREDIENTS TO MAKE 2 CUPS

1 litre (1¾ pints) kettle-hot water
2 tsp Oolong tea leaves
6 fresh chamomile flowers
6 fresh peppermint leaves
6 fresh strawberries, hulled and coarsely chopped
1 tsp clear honey
Ice cubes, fresh mint leaves and cucumber slices, to serve

This is a refreshing and delicious summertime drink that can be created the day before and chilled in the fridge overnight.

Oolong is made from *Camellia sinensis*, or China tea. It is a partially fermented loose leaf tea which is high in polyphenol antioxidants that have long been known to benefit the skin. The restorative powers of chamomile and peppermint help to create a summer herbal infusion that will keep you and your skin hydrated and cool.

Strawberries are also rich in antioxidants, including ellagic acid and anthocyanins, which are anti-inflammatory and help reduce skin damage from the sun's UV rays by inhibiting certain enzymes that contribute to the breakdown of collagen. Another is salicylic acid, a beta hydroxy acid, that can help lessen the appearance of hyperpigmentation or age spots.

Use 500ml (18fl oz) of the hot water to warm the teapot and cups. Pour the water away.

Add the Oolong tea leaves, chamomile and peppermint to the warmed pot. Pour in the remaining 500ml (18fl oz) of hot water and stir.

Infuse the tea for 3–4 minutes, then strain into a jug. Add the chopped strawberries and honey, stir and seal with a lid or a clean cloth, cool and refrigerate the tea overnight.

Serve the herbal tea cold, with ice, mint leaves and cucumber slices.

AUTUMN SKIN FORMULATIONS

—

CRAB APPLE & RED CLOVER SPOT BREAKOUT TREATMENT

INGREDIENTS TO MAKE 250ML (9FL OZ) BOTTLE

- 20g (¾oz) fresh red clover flowers
- 200ml (7fl oz) apple or crab apple cider vinegar (see page 120)
- 2 tbsp witch hazel
- 2 tbsp rosewater

The anti-inflammatory properties of red clover have calming effects for sufferers of acne. Breakouts can be managed with attention to gentle hygiene, plenty of hydration, staying out of the sun, wearing chemical-free cosmetics and avoiding touching the affected skin.

Apple and crab apple cider vinegars' anti-inflammatory properties are derived from the alpha hydroxy acid it contains, helping to exfoliate the skin, absorb excess oils and unblock pores. This acid also has the power to restore the normal, slightly acidic pH levels of our skin. You should follow the apple and crab apple cider vinegar making method outlined in the introduction to this chapter (see page 120) before starting.

Place the red clover flowers in a sterilized glass jar and fill with the apple cider vinegar (making sure the flowers are covered). Seal, label and date the jar and leave in a cool, dark cupboard for 4 weeks.

Strain and decant the liquid into a clean jug and mix in the witch hazel and rosewater.

Decant the mixture into a sterilized bottle, seal, label and date it. It will keep in the fridge for up to 3 months. Use topically on breakouts.

AUTUMN HERBAL TEA - GERANIUM, CLOVER, CALENDULA

INGREDIENTS FOR 2 CUPS
1 litre (1¾ pints) kettle-hot water
2 tsp honeybush tea leaves
6 fresh calendula flowers
6 fresh scented pelargonium leaves and flowers
6 fresh red clover flowers
2 slices of fresh pear
2 slices of fresh plum
1 vanilla pod, split

In an 'Indian summer', autumn days can still be filled with warm sunshine and flower borders are still in bloom. I harvest scented pelargonium, red clover and calendula petals well into mid-late autumn.

The fermented leaves and stems of *Cyclopia intermedia*, a type of legume indigenous to South Africa, are used to brew honeybush tea, which is similar to rooibos tea. The honeybush grows wild only in a small area of South Africa's Western and Eastern Cape provinces, but it is now cultivated to make this aromatic herbal tea. Honeybush extract and fermented honeybush tea have been reported to inhibit the formation and appearance of wrinkles caused by UV irradiation – so called photoageing – prevent thickening of the epidermal layer, and suppress collagen tissue breakdown, making it invaluable as a natural anti-ageing ingredient.

Here, it is paired with orchard fruit and vanilla, both of which are rich in antioxidants that are known to neutralize free radicals (see page 19) and reverse the skin damage they cause. The vanilla pod will also add sweetness to the brew without the added calories of sugar.

Enjoy cups of this colourful and delicious tea to keep you warm and hydrated while gardening; the more active and resilient you remain in the lead-up to winter, the better for your health and well-being.

Use 500ml (18fl oz) of the hot water to warm the teapot and cups. Pour the water away.

Add the tea leaves, the calendula, pelargonium and red clover flowers and leaves, the fruits and vanilla pod to the warmed pot. Pour in the remaining 500ml (18fl oz) of hot water and stir.

Infuse the tea for 3–4 minutes, then strain into warm cups to serve.

AUTUMN FACIAL SPA

For the Rosehip Facial Toner

INGREDIENTS TO MAKE A 250ML (9FL OZ) BOTTLE
- 100ml (3½fl oz) rosewater
- 100ml (3½fl oz) distilled water
- 30g (1oz) rosehips, coarsely chopped
- 20 drops of rosehip tincture (see page 119)
- 10 drops of rose absolute extract

For the Rosehip Sunshine Facial Serum

INGREDIENTS TO MAKE A 250ML (9FL OZ) BOTTLE
- 200g (7oz) dried rosehips, coarsely chopped
- 50ml (2fl oz) rosehip macerated oil (see page 109)
- 50ml (2fl oz) apricot kernel oil
- 50ml (2fl oz) hemp oil
- 50ml (2fl oz) borage seed oil

After the heat of summer and before the central heating goes on in winter, it's restorative to give your skin a healing and rejuvenating rosehip skin spa in a few easy steps. This routine can be done at any time of the day on cleansed skin, but it is particularly beneficial at night time, accompanied by a herbal tisane and a warm bath.

A handmade, botanical facial serum is an oil- or water-based treatment which delivers highly concentrated, targeted nutrients to help resolve one or several skin problems. Here, I use rosehips, which are high in vitamin C, Omega-3 and 6, and are widely used in herbal medicine and natural skincare formulations to boost the immune system and to help delay the signs of skin ageing, by helping to regenerate cellular membranes and tissues.

Step 1 Cleanse your skin with a botanical soap (see page 111) and rinse with warm water infused with coarsely chopped fresh rosehips for 10 minutes. Gently pat your face dry with a clean towel.
Step 2 Using a warm, damp linen cloth or facial brush, gently exfoliate the face and neck using circular motions, away from the heart.
Step 3 Using reusable cotton or linen pads, apply some rosehip facial toner to your face and neck with a circular motion.
Step 4 Finally, massage in a little rosehip sunshine facial serum, gently patting your skin with your fingertips and using circular motions. Any excess oil can be wiped on your hands and elbows.

For the Rosehip Facial Toner

Place all the ingredients in a sterilized jar, seal and shake. Label and place in a cool, dark cupboard for 2 weeks.

Strain and decant into a sterilized bottle. Seal, label and date. Refrigerate and use within 3 months.

For the Rosehip Sunshine Facial Serum

Place the rosehips in a sterilized jar and completely cover with the oils. Seal and shake. Label and date the jar and place on a sunny window sill for 1–2 weeks to extract the active properties of the rosehips. Shake daily.

Strain and decant into a sterilized bottle. Seal, label and date. Use as part of your facial spa, following the steps opposite. The serum will keep for 2–3 weeks in a cool, dark place.

AUTUMN HARVEST HYDROSOL TONIC & BODY BUTTER

Autumn Harvest Hydrosol Tonic

INGREDIENTS
1 litre (1¾ pints) water
The second tier of the steamer pot filled with the aerial parts of autumn herbs and flowers (there will probably be some rosehips, crab apples and foraged herbs as well)
Ice cubes

Autumn Harvest Body Butter
INGREDIENTS TO MAKE A 200ML (7FL OZ) JAR
2 tbsp cacao butter
2 tbsp coconut oil
2 tbsp shea butter
2 tbsp evening primrose oil
50g (1¾oz) autumn harvest fresh flowers, herbs, leaves and berries, grown in the skincare garden or foraged
6 drops of lavender essential oil
6 drops of geranium essential oil

At the end of autumn, when you are starting to cut back parts of the garden after harvest, there are always some flowers and herbs still in bloom and blush. These are the last of the garden's floral display and I hold on to it for colour and pleasure, but it's also useful for my end-of-year hydrosol tonic and body butter. I snip at rosehips and the last rose petals, chamomile, evening primrose, dandelion and borage flowers. In the fields, I forage for crab apples and fill baskets with wild flowers and herbs.

Come the cold, dark days of winter, these products evoke memories of summer's sunshine and scent. I use my autumn harvest hydrosol as a facial toner, in tisanes, in the bath, on a cloth as a refresher and also as a facial or pillow mist. The body butter is a winter mainstay for dry hands, elbows and the neck, as well as an all-over treatment after dry skin brushing and bathing.

You should follow the hydrosol and enfleurage making methods outlined on pages 115 and 116 before starting.

For the Autumn Harvest Hydrosol Tonic

Pour the water in to the bottom pan, and place the plant material in the second tier or steamer basket above it, ensuring that the water does not reach the base of the middle tier where your plant material sits. Make sure that this tier is quite full, leaving just 1cm (½in) free space at the top.

Select a heatproof bowl slightly smaller than the size of the steamer, one that will fit in the top tier, which is where the distillate will condense. On top of this tier, place an inverted lid topped with lots of ice cubes, or ice bags. When the rising steam distillate meets the ice barrier, it will condense back to liquid and flow down into the bowl beneath.